Oil Sand Production Processes

Oil Sand Production Processes

James G. Speight, PhD, DSc

CD & W Inc., Laramie, Wyoming, USA

AMSTERDAM • BOSTON • HEIDELBERG • LONDON
NEW YORK • OXFORD • PARIS • SAN DIEGO
SAN FRANCISCO • SINGAPORE • SYDNEY • TOKYO
Gulf Professional Publishing is an imprint of Elsevier

Gulf Professional Publishing is an imprint of Elsevier
The Boulevard, Langford Lane, Kidlington, Oxford, OX5 1GB, UK
225 Wyman Street, Waltham, MA 02451, USA

First published 2013

British Library Cataloguing in Publication Data
A catalogue record for this book is available from the British Library

Library of Congress Cataloging-in-Publication Data
A catalog record for this book is available from the Library of Congress

ISBN: 978-0-12-404572-9

For information on all Gulf Professional Publishing
publications visit our website at **store.elsevier.com**

This book has been manufactured using Print On Demand technology. Each copy is produced to
order and is limited to black ink. The online version of this book will show color figures where
appropriate.

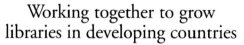

Working together to grow
libraries in developing countries

www.elsevier.com | www.bookaid.org | www.sabre.org

ELSEVIER   BOOK AID
            International   Sabre Foundation

Transferred to Digital Printing in 2012

TABLE OF CONTENTS

The declining reserves of light crude oil have resulted in an increasing need to develop options to upgrade the abundant supply of known reserves of oil sand bitumen. In addition, there is considerable focus and renewed efforts on adapting recovery techniques to the production of bitumen from oil sand deposits.

While information is included related to worldwide oil sand deposits and bitumen, the majority of work performed on oil sands has focused on the oil sands of Alberta, Canada, the majority of the work described in this text focuses on the Canadian oil sands.

The overall objectives of this book are to present to the reader the current methods of recovery for bitumen by nonthermal and thermal methods. The book commences with the operative definitions followed by the properties of oil sand and oil sand bitumen, since both sets of properties have an effect on bitumen recovery. Then follow chapters in which the various methods of bitumen recovery are described. Since bitumen, in all senses of the word, is a poor refinery feedstock, a chapter is also devoted to a discussion of the potential and benefits of bitumen upgrading during *in situ* recovery. Finally, a chapter is also devoted to the various environmental aspects of oil sand technology.

This book is designed to be suitable for undergraduate students, graduate students, and professionals who are working with heavy oil, extra heavy oil, and oil sand bitumen.

Each chapter also includes a list of references that will serve to guide the reader to the source of more detailed information.

James G. Speight
Laramie, WY

LIST OF FIGURES

LIST OF TABLES

Definitions

1.1 INTRODUCTION

With the potential energy shortages facing many countries, it is imperative that attention be focused on new potential reserves for alternate energy supplies (Lee, 1996; Lee et al., 2007; Speight, 2009). There are many possible sources (such as biomass) available but the various oil sand (tar-sand) deposits offer release from the energy crunch and dependence upon foreign sources of petroleum.

With such an incentive as an assured supply of energy, the time and effort required to achieve the required production is worthwhile. The challenge is to develop the technology to produce enough oil at a reasonable market value and, at the same time, conserve the environment. This requires development of more sophisticated extraction procedures and optimizing upgrading processes to give maximum yields of usable product from the bitumen recovered.

However, before progressing into more detailed aspects of oil sand processing, it is necessary to understand the nature of the materials that derive from the petroleum heavy oil family and the oil sand bitumen family. This is best achieved by considering the definitions of these materials to ensure that the reader understands, and is in accord with, the prevalent methods of recovery, properties, and processing.

As a means in introduction, in any text related to the properties and behavior (recovery or reining) of a natural resource (such as oil sand), it is necessary to understand the resource first through the name or terminology or definition (Smalley, 2000; Speight, 2007, 2009).

1.2 DEFINITIONS

Definitions are the means by which scientists and engineers communicate the nature of a material to each other and to the world, through either the spoken or the written word. On the other hand, *terminology* is the means by which various subjects are named so that reference can be made in conversations and in writings and so that the meaning is passed on. While terminology might be used for a general understating of a subject area or a material, it is not always an accurate description of the subject area or material under discussion—this is accomplished by use of relevant and meaningful definitions.

Particularly troublesome in this respect, and more confusing, are those terms that are applied to the more viscous materials, for example, the use of the terms *bitumen* and *asphalt*—both of which are claimed to be present in oil sand. This part of the text attempts to alleviate much of the confusion that exists, but it must be remembered that terminology is still open to personal choice and historical usage and for the purposes of this text, the word *bitumen* is used to mean the organic material that is present in oil sand deposits. It is through such definitions that differences can be ascribed to petroleum vis-à-vis heavy oil vis-à-vis oil sand bitumen.

In spite of many efforts, a uniform definition of the term unconventional oil has not been generally accepted. The pragmatic reason for the differentiation between conventional oil and unconventional oil is the greater technical effort and expenditure for extracting unconventional oil. Unconventional oil comprises bitumen from oil sands, as well as extra heavy oil and crude oil from coal and/or oil shale. Thus the denomination *unconventional* can refer to (1) geological aspects of the formation, (2) properties of the deposits, and (3) technical necessities for an ecologically acceptable, economic exploitation, and more appropriately to the method of recovery.

In addition to the definitions of petroleum, heavy oil, and oil sand bitumen presented below, there are two definitions that need to be addressed which also speak to the difference between petroleum and heavy oil on the one hand and oil sand bitumen on the other. These are (1) reservoir and (2) deposit and will be presented first in order to affirm the differences between petroleum/heavy oil and oil sand bitumen.

1.2.1 Reservoir

Petroleum is derived from aquatic plants and animals that lived and died hundreds of millions of years ago. Their remains became mixed with mud and sand in layered deposits that, over the millennia, were geologically transformed into sedimentary rock. Gradually the organic matter decomposed and eventually formed petroleum (or a related precursor), which migrated from the original source beds to more porous and permeable rocks, such as *sandstone* and *siltstone*, where it finally became entrapped. Such entrapped accumulations of petroleum are called *reservoirs*. A series of reservoirs within a common rock structure or a series of reservoirs in separate but neighboring formations is commonly referred to as an *oil field*. A group of fields is often found in a single geologic environment known as a *sedimentary basin* or *province*.

Simply, a reservoir is a trap in which petroleum and heavy oil are contained. A trap forms when there is a sealing medium which prevents the oil from escaping to a lower or upper formation. In short, there is an impermeable lower formation (basement rock) and an impermeable upper formation (cap rock).

Traps are broadly classified into three categories that are based on geological characteristics: (1) structural trap, (2) stratigraphic trap, and (3) hydrodynamic trap. The trapping mechanisms for many petroleum reservoirs have characteristics from several categories and can be known as a combination trap.

Structural traps are formed as a result of changes in the structure of the subsurface due to processes such as folding and faulting, leading to the formation of anticline, folds, and domes. They are more easily delineated and more prospective than their stratigraphic counterparts, with the majority of the world's petroleum reserves being found in structural traps. *Stratigraphic traps* are formed as a result of lateral and vertical variations in the thickness, texture, porosity, or lithology of the reservoir rock. Examples of this type of trap are an unconformity trap, a lens trap, and a reef trap. Hydrodynamic traps are a less common type of trap than the structural traps and stratigraphic traps and are caused by the differences in water pressure that are associated with water flow, creating a tilt of the hydrocarbon–water contact.

Typically a petroleum trap contains fluids (liquids and gases) that are mobile under reservoir conditions of temperature and pressure. Furthermore, the fluids can be produced under the conditions of existing reservoir energy (primary recovery methods) and by using conventional secondary recovery methods (such as waterflood) or by using enhanced recovery methods (such as steam flood) (Speight, 2007, 2008, 2009).

1.2.2 Deposit

The term *deposit* is typically applied to an accumulation of ore in which the ore is part of the rock.

Oil sand (tar-sand) deposits (or, more technically, bituminous sand deposits) are loose sand or partially consolidated sandstone containing naturally occurring mixtures of sand, clay, and water, saturated with a dense and extremely viscous hydrocarbonaceous material technically referred to as *bitumen* (or colloquially as *tar* due to its similar appearance, odor, and color to tar produced thermally from coal). The bitumen contained in oil sand deposits exists in the semisolid or solid phase and is typically immobile under deposit conditions of temperature and pressure, unless heated or diluted with low boiling hydrocarbon solvents. Attempts to define oil sand bitumen based on a single property such as API gravity or viscosity are, at best, speculative and subject to inaccuracies.

It is preferable to define bitumen in an oil sand deposit as the hydrocarbonaceous material *which is not recoverable in its natural state by conventional oil well production methods including currently used enhanced recovery techniques.*

The majority of information available for oil sand deposits has become available through the extensive work carried out on the Athabasca (Alberta, Canada) deposits. The Alberta deposits have been evaluated by the drilling of numerous wells, by extensive laboratory and field experimental work directed toward development of both mining and *in situ* methods, and by the work at Athabasca on a commercial scale.

The oil sands are composed of a series of quartz sand deposits impregnated with varying amounts of heavy, highly viscous bitumen. The sands are deposited on an irregular surface of limestone or shale and are generally overlain with overburden, varying in thickness from 0 to 2,000 ft. The bitumen has an API gravity of approximately 7–9° and contains approximately 5% w/w sulfur and is relatively high in metals (particularly nickel and vanadium). The deposit is highly compacted and temperature within the ore body (bitumen-containing sand) remains fairly constant at approximately 6°C (43°F). At this temperature, the bitumen has a viscosity so high that it is immobile (Chapter 2).

One of the major concerns in any process that may be utilized in developing oil sand resources is the effect on the environment. Mining processes (Chapter 3) disturb large areas that must be rehabilitated and restored to acceptable condition. The *in situ* recovery process (Chapters 4 and 5) also disturbs some land area but considerably less so than mining processes; nonetheless restoration must be considered. Both methods run the risk of water contamination, and provision must be made to assure that pollution is minimized.

Upgrading operations from both methods can produce air pollution through sulfur and particulates emissions to the atmosphere so that new plants can remain within the strict environmental standards.

With respect to Canadian efforts, the development of the oil sand industry has begun. The Federal Government of Canada and the Provincial Governments of Alberta have realized that this is a challenge worth facing and have met the challenge by showing active support to develop the needed technology. With combined government/industry effort, progress should be more quickly achieved and some of the problems solved.

1.2.3 Petroleum

Petroleum (*crude oil, conventional crude oil*) is found in the microscopic pores of sedimentary rocks, such as sandstone and limestone (Speight, 2007). Not all of the pores in a rock contain petroleum and some pores will be filled with water or brine that is saturated with minerals. However, not all of the oil fields that are discovered are exploited since the oil may be far too deep or of insufficient volume or the oil field may be so remote that transport costs would be excessively high.

The *definition* of petroleum (and the equivalent term *crude oil*) has been varied, unsystematic, diverse, and often archaic. In fact, there has been a tendency to define petroleum and heavy oil on the basis of a single property. While this may be suitable for a general understanding, it is by no means accurate and does not reflect the true nature of petroleum or heavy oil or the characterization of the material. Unfortunately, this form of *identification* or differentiation is a product of many years of growth and its long established use, however general or inadequate it may be, is altered with difficulty, and a new term, however precise, is adopted only slowly.

Petroleum is a naturally occurring mixture of hydrocarbons, generally in a liquid state, which may also include compounds of sulfur nitrogen oxygen metals and other elements (ASTM D4175).

Thus, the term *petroleum* covers a wide assortment of materials consisting of mixtures of hydrocarbons and other compounds containing variable amounts of sulfur, nitrogen, and oxygen, which may vary widely in specific gravity, API gravity, and the amount of residuum (Table 1.1). Metal-containing constituents, notably those compounds that contain vanadium and nickel, usually occur in the more viscous crude oils in amounts up to several thousand parts per million and failure to control their levels can have serious consequences during processing of these feedstocks (Speight, 1984, 2007). Because petroleum is a mixture of

Table 1.1 Typical Variations in the Properties of Petroleum			
Petroleum	Specific Gravity	API Gravity	Residuum >1,000°F, % v/v
US domestic			
California	0.858	33.4	23.0
Oklahoma	0.816	41.9	20.0
Pennsylvania	0.800	45.4	2.0
Texas	0.827	39.6	15.0
Texas	0.864	32.3	27.9
Foreign			
Bahrain	0.861	32.8	26.4
Iran	0.836	37.8	20.8
Iraq	0.844	36.2	23.8
Kuwait	0.860	33.0	31.9
Saudi Arabia	0.840	37.0	27.5
Venezuela	0.950	17.4	33.6

widely varying constituents and proportions, its physical properties also vary widely and the color varies from near colorless to black.

In the crude state, petroleum has minimal value, but when refined it provides high-value liquid fuels, solvents, lubricants, and many other products. Crude petroleum can be separated into a variety of different generic fractions by distillation. And the terminology of these fractions has been bound by utility and often bears little relationship to composition.

The fuels derived from petroleum contribute approximately one-third to one-half of the total world energy supply and are used not only as transportation fuels (i.e., gasoline, diesel fuel, and aviation fuel, among others) but also to heat buildings. Petroleum products have a wide variety of uses that vary from gaseous and liquid fuels to near-solid machinery lubricants. In addition, the residue of many refinery processes, asphalt—a once-maligned by-product, is now a premium value product for highway surfaces, roofing materials, and miscellaneous waterproofing uses.

The molecular boundaries of petroleum cover a wide range of boiling points and carbon numbers of hydrocarbon compounds and other compounds containing nitrogen, oxygen, and sulfur, as well as metal-containing (porphyrin) constituents. However, the actual boundaries of such a *petroleum map* can only be arbitrarily defined in terms of boiling point and carbon number. In fact, petroleum is so diverse that materials from different sources exhibit different boundary limits, and for this reason alone it is not surprising that petroleum has been difficult to *map* in a precise manner.

Since there is a wide variation in the properties of crude petroleum, the proportions in which the different constituents occur vary with the origin and the relative amounts of the source materials that form the initial *protopetroleum* as well as the maturation conditions. Thus, some crude oils have higher proportions of the lower boiling components and others (such as heavy oil and bitumen) have higher proportions of higher boiling components (asphaltic components and residuum).

Petroleum occurs underground, at various pressures depending on the depth. Because of the pressure, it contains considerable natural gas in solution. Petroleum underground is much more fluid than it is on the surface and is generally mobile under reservoir conditions because the elevated temperatures (the *geothermal gradient*) in subterranean formations decrease the viscosity. Although the geothermal gradient varies from place to place, it is generally on the order of 8°C per 1,000 ft of depth (15°F per 1,000 ft of depth) or 0.008°C per foot of depth (0.015°C per foot of depth).

1.2.4 Heavy Oil

There are large resources of *heavy oil* in Canada, Venezuela, Russia, the United States, and many other countries. The resources in North America alone provide a small percentage of current oil production (approximately 2%); existing commercial technologies could allow for significantly increased production. Under current economic conditions, heavy oil can be profitably produced, but at a smaller profit margin than for conventional oil, due to higher production costs and upgrading costs in conjunction with the lower market price for heavier crude oils.

Heavy oil is a type of petroleum that is viscous and contains a higher level of sulfur than conventional petroleum occurring in similar locations to petroleum (Ancheyta and Speight, 2007; IEA, 2005; Speight, 2007, 2009). The nature of heavy oil is a problem for recovery operations and for refining—the viscosity of the oil may be too high, thereby rendering recovery expensive and/or difficult and the sulfur content may be high, which increases the cost of refining the oil.

The name *heavy oil* can often be misleading as it has also been used in reference to (1) fuel oil that contains residuum left over from distillation, that is, residual fuel oil, (2) coal tar creosote, or (3) viscous crude oil. For most purposes of this text, the term heavy oil is used to mean *viscous crude oil*, which is a petroleum-type material but which requires more energy (through the application of secondary or tertiary recovery methods) to extract it from the reservoir (Speight, 2009).

Thus, heavy oil is a *type* of petroleum that is different from conventional petroleum insofar as it is much more difficult to recover from the subsurface reservoir. Such material has a much higher viscosity (and lower API gravity) than conventional petroleum, and recovery of petroleum types such as this usually requires thermal stimulation of the reservoir. When petroleum occurs in a reservoir that allows the crude material to be recovered by pumping operations as a free-flowing dark to light colored liquid, it is often referred to as *conventional petroleum*.

Very simply, heavy oil is a type of crude oil which is very viscous and does not flow easily. The common characteristic properties (relative to conventional crude oil) are high specific gravity, low hydrogen to carbon ratios, high carbon residues, and high contents of asphaltene constituents, heavy metal, sulfur, and nitrogen. Specialized recovery and refining processes are required to produce more useful fractions, such as naphtha, kerosene, and gas oil.

Heavy oil is an oil resource that is characterized by high viscosities (i.e., resistance to flow) and high densities compared to conventional oil. Most heavy oil reservoirs originated as conventional oil that formed in deep formations, but migrated to the surface region where they were degraded by bacteria and by weathering, and where the lightest hydrocarbons escaped. Heavy oil is deficient in hydrogen and has high carbon, sulfur, and heavy metal content. Hence, heavy oil requires additional processing (upgrading) to become a suitable refinery feedstock for a normal refinery.

Heavy oil accounts for more than double the resources of conventional oil in the world and heavy oil offers the potential to satisfy current and future oil demand. Not surprisingly, heavy oil has become an important theme in the petroleum industry with an increasing number of operators getting involved or expanding their plans in this market around the world.

However, as said, heavy oil is more difficult to recover from the subsurface reservoir than conventional or light oil. A very general definition of heavy oils has been, and remains, based on the API gravity or viscosity, and the definition is quite arbitrary although there have been attempts to rationalize the definition based upon viscosity, API gravity, and density.

For example, heavy oils were considered to be those crude oils that had gravity somewhat less than 20° API with the heavy oils falling into the API gravity range 10–15°. For instance, Cold Lake heavy crude oil has an API gravity equal to 12° and extra heavy oils, such as oil sand bitumen, usually have an API gravity in the range 5–10° (Athabasca bitumen = 8° API). However, extra heavy oil has flow characteristics, with mobility in the reservoir, that the bitumen in the deposit does not have (Attanasi and Meyer, 2007; Meyer et al., 2007; Babies and Messner, 2012). Residua would vary depending upon the temperature at which distillation was terminated but usually vacuum residua are in the range 2–8° API (Speight, 2000 and references cited therein; Speight and Ozum, 2002 and references cited therein).

Heavy oil has a much higher viscosity (and lower API gravity) than conventional petroleum and recovery of heavy oil usually requires thermal stimulation of the reservoir. The generic term *heavy oil* is often applied to a crude oil that has less than 20° API and usually, but not always, sulfur content is higher than 2% by weight (Speight, 2000). Furthermore, in contrast to conventional crude oils, heavy oils are darker in color and may even be black.

The term *heavy oil* has also been arbitrarily (incorrectly) used to describe both the heavy oils that require thermal stimulation of recovery from the reservoir and the bitumen in bituminous sand (oil sand) formations from which the heavy bituminous material is recovered by a mining operation. *Extra heavy oil* is a nondescript term (related to viscosity) of little scientific meaning which is usually applied to oil sand bitumen, which is generally incapable of free flow under reservoir conditions.

The methods outlined in this book for heavy oil recovery focus on heavy oil with an API gravity of less than 20° and examples of such heavy oils are presented (Table 1.2). However, it must be recognized that

Table 1.2 API Gravity and Sulfur Content of Selected Heavy Oils		
	API	**Sulfur (wt%)**
Bachaquero	13.0	2.6
Boscan	10.1	5.5
Cold Lake	13.2	4.1
Huntington Beach	19.4	2.0
Kern River	13.3	1.1
Lagunillas	17.0	2.2
Lloydminster	16.0	2.6
Lost Hills	18.4	1.0
Merey	18.0	2.3
Midway Sunset	12.6	1.6
Monterey	12.2	2.3
Morichal	11.7	2.7
Mount Poso	16.0	0.7
Pilon	13.8	1.9
San Ardo	12.2	2.3
Tremblador	19.0	0.8
Tia Juana	12.1	2.7
Wilmington	17.1	1.7
Zuata Sweet	15.7	2.7
For reference, Athabasca tar-sand bitumen has API = 8° and sulfur content = 4.8% w/w.		

some of these heavy oils are pumpable and are already being recovered by this method. Recovery depends not only on the characteristics of the oil but also on the characteristics of the reservoir—including the temperature of the reservoir and the pour point of the oil (Chapter 2). Oil sand bitumen falls into a range of high viscosity (Figure 1.1) and the viscosity is subject to temperature effects (Speight, 2007, 2009), which is the reason for the application of thermal methods to heavy oil recovery.

1.2.5 Oil Sand Bitumen

Oil sand bitumen, on the other hand, is different to petroleum and heavy oil. The term *bitumen* (also, on occasion, incorrectly referred to as *native asphalt* and *extra heavy oil*) includes a wide variety of reddish brown to black materials of semisolid, viscous to brittle character that can exist in nature with no mineral impurity or with mineral matter contents that exceed 50% by weight. Bitumen is frequently found filling pores and crevices of sandstone, limestone, or argillaceous sediments, in which case the organic and associated mineral matrix is known as *rock asphalt* (Abraham, 1945).

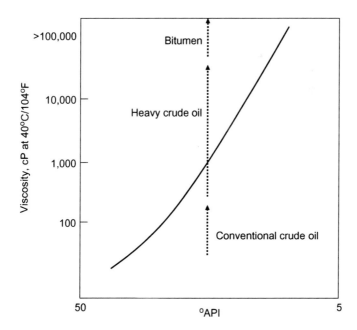

Fig. 1.1 General relationship of viscosity to API gravity.

Bitumen is a naturally occurring material that is found in deposits where the permeability is low and passage of fluids through the deposit can only be achieved by prior application of fracturing techniques. Oil sand bitumen is a high boiling material with little, if any, material boiling below 350°C (660°F) and the boiling range is approximately the same as the boiling range of an atmospheric residuum.

Display Quote

Tar sands have been defined in the United States (FE-76-4) as:

… the several rock types that contain an extremely viscous hydrocarbon which is not recoverable in its natural state by conventional oil well production methods including currently used enhanced recovery techniques. The hydrocarbon-bearing rocks are variously known as bitumen-rocks oil, impregnated rocks, oil sands, and rock asphalt.

Inference petroleum and heavy oil are recoverable by well production methods and currently used enhanced recovery techniques. For convenience, it is assumed that before depletion of the reservoir energy, conventional crude oil is produced by primary and secondary techniques while heavy oil requires tertiary (enhanced) oil recovery (EOR) techniques. While this is an oversimplification, it may be used as a general guide.

The term *natural state* cannot be defined out of context and in the context of FEA Ruling 1976-4 the term is defined in terms of the composition of the heavy oil or bitumen. The final determinant of whether or not a reservoir is an oil sand deposit is the character of the viscous phase (bitumen) and the method that is required for recovery.

Generally, bitumen is solid or near solid at room temperature and is solid or near solid at reservoir temperature. In other words, oil sand bitumen is immobile in the reservoir and requires conversion or extreme stimulation for recovery.

Generally, the bitumen found in oil sand deposits is an extremely viscous material that is *immobile under reservoir conditions* and cannot be recovered through a well by the application of secondary or enhanced recovery techniques.

Thus, the expression *oil sand* is commonly used in the petroleum industry to describe sandstone reservoirs that are impregnated with a heavy, viscous black crude oil that cannot be retrieved through a well by conventional production techniques (FE-76-4, above). However, the terms *oil sand* and tar sand are misnomers—the sands do not contain oil or tar. The name *tar* is usually applied to the heavy product remaining after the destructive distillation of coal or other organic matter (Speight, 2013). Current recovery operations of bitumen in oil sand formations are predominantly focused on a mining technique.

Thus, alternative names, such as *bituminous sand* or *oil sand*, are gradually finding usage, with the former name (bituminous sands) more technically correct. The term *oil sand* is also used in the same way as the term *tar sand*, and these terms are used interchangeably throughout the industry.

Bituminous rock and *bituminous sand* are those formations in which the bituminous material is found as a filling in veins and fissures in fractured rocks or impregnating relatively shallow sand, sandstone, and limestone strata. These terms are, in fact, the more correct geological description of *oil sand*. The deposits contain as much as 20% of bituminous material, and if the organic material in the rock matrix is bitumen, it is usual (although chemically incorrect) to refer to the deposit as *rock asphalt* to distinguish it from bitumen that is relatively mineral free. A standard test (ASTM D4) is available for determining the bitumen content of various mixtures with inorganic materials, although the use of the word *bitumen* as applied in this test might be questioned and it might be more appropriate to use the term *organic residues* to include *tar* and *pitch*. If the material is of the asphaltite type or asphaltoid type, the corresponding terms should be used: rock asphaltite or rock asphaltoid.

Bituminous rocks generally have a coarse, porous structure, with the bituminous material in the voids. A much more common situation is that the organic material is present as an inherent part of the rock composition insofar as it is a diagenetic residue of the organic material detritus that was deposited with the sediment. The organic components of such rocks are usually refractory and are only slightly affected by most organic solvents.

A special class of bituminous rocks that has achieved some importance is the so-called *oil shale*. These are argillaceous, laminated sediments of generally high organic content that can be thermally decomposed to yield appreciable amounts of oil, commonly referred to as *shale oil*. Oil shale does not yield shale oil without the application of high temperatures and the ensuing thermal decomposition that is necessary to decompose the organic material (*kerogen*) in the shale.

Sapropel is an unconsolidated sedimentary deposit rich in bituminous substances. It is distinguished from peat in being rich in fatty and waxy substances and poor in cellulosic material. When consolidated into rock, sapropel becomes oil shale, bituminous shale, or boghead coal. The principal components are certain types of algae that are rich in fats and waxes. Minor constituents are mineral grains and decomposed fragments of spores, fungi, and bacteria. The organic materials accumulate in water under reducing conditions.

Although the term *tar sand* is in common use, it is incorrect to refer to native bituminous materials as *tar* or *pitch*. The word tar might indeed be descriptive of the black, heavy bituminous material; it is best to avoid its use with respect to natural materials and to restrict the meaning to the volatile or near-volatile products produced in the destructive distillation of such organic substances as coal (Speight, 2013). In the simplest sense, pitch is the distillation residue of the various types of tar.

Finally, *asphalt* is a product of petroleum refining and is the viscous product that finds a variety of uses in, for example, road paving, roofing, and asphalt paints. To refer to oil sand bitumen as natural asphalt is a common but incorrect and misleading use of the word asphalt. Many countries throughout the world use the word *bitumen* for *road asphalt* but such usage is equally incorrect and misleading.

1.3 RECOVERY OF PETROLEUM, HEAVY OIL, AND OIL SAND BITUMEN

Recovery, as applied in the petroleum industry, is the production of petroleum or heavy oil from a reservoir and oil sand bitumen from a deposit (Speight, 2007, 2009).

There are several methods by which recovery of petroleum or heavy oil can be achieved that range from recovery due to reservoir energy (i.e., the oil flows from the well hole without assistance) to enhanced recovery methods in which considerable energy must be added to the reservoir to produce the oil. However, the effect of the method on the oil and on the reservoir must be considered before application.

Conventional petroleum production can include up to three distinct phases: primary, secondary, and tertiary (or enhanced) recovery. During primary recovery, the natural pressure of the reservoir, or gravity, drives oil into the wellbore, which, when combined with artificial lift techniques (such as pumps), brings the oil to the surface.

Primary recovery occurs as a result of the generation of natural energy from expansion of gas and water within the producing formation, pushing fluids into the wellbore and lifting them to the surface. Typically, approximately 10% v/v of the original oil in place in the reservoir is produced during primary recovery.

Secondary recovery occurs as artificial energy is applied to lift fluids to the surface. This may be accomplished by injecting gas down a hole to lift fluids to the surface, installation of a subsurface pump, or injecting gas or water into the formation itself. Secondary recovery is effected when well, reservoir, facility, and economic conditions permit. Secondary recovery methods result in the recovery of an additional 20–40% v/v of the original oil in place.

Tertiary recovery methods (*enhanced recovery methods, EOR methods*) are applied when the means of increasing fluid mobility in oil reservoirs within the reservoir are introduced in addition to secondary techniques. This may be accomplished by introducing additional heat into the formation (such as by the injection of steam—*steam flood*) to lower the viscosity, i.e., thin the oil, and improve its ability to flow to the wellbore.

Tertiary recovery technique involves injecting bacteria into the oil field. Some bacteria produce polysaccharides which reduce the permeability of the water-filled pores of the reservoir rock and this effectively forces injected water into the oil-filled pores, pushing the oil out. Other bacteria produce carbon dioxide which helps to increase pressure within the rock pores, forcing out the oil. Other bacteria produce surfactants and/or chemicals that reduce the viscosity of the oil.

On the other hand, proposed methods for recovery of bitumen from oil sand deposits are based on either (1) mining or (2) advanced *in situ* processes (beyond those processes accepted as conventional or enhanced recovery methods) combined with some further processing or operation on the oil sands *in situ*. The typical *in situ* recovery methods are not applicable to bitumen recovery because bitumen, in its immobile state, is extremely difficult to move to a production well.

Extreme processes are required, usually in the form of a degree of thermal conversion that produces free-flowing product oil that will flow to the well and reduce the resistance of the bitumen to flow. Oil sand deposits are not amenable to injection technologies such as steam soak and steam flooding.

In fact, the only successful commercial method of recovering bitumen from oil sand deposits is used at the oil sand plants in Alberta (Canada) and involves use of a mining technique and enhanced recovery methods that fall outside of the specified enhanced recovery methods used for conventional petroleum and heavy oil.

1.4 RATIONALIZATION OF THE DEFINITIONS

The need for understandable definitions is obvious in the light of the confusing terminology that has been used since the dawn of the modern petroleum industry.

For example, there has been an attempt to classify petroleum, heavy oil, and oil sand bitumen using the viscosity scale with 10,000 cP being the fine line of demarcation between heavy oil and oil sand bitumen (Speight, 2007, 2009).

Use of such a system leads to confusion when having to differentiate between a material having a viscosity of 9,950 cP and one having a viscosity of 10,050 cP as well as taking into account the limits of accuracy of the method of viscosity determination. Whether the limits are the usual laboratory experimental difference ($\pm 3\%$) or more likely the limits of accuracy of the method ($\pm 5\%$ to $\pm 10\%$), there is the question of accuracy when tax credits for recovery of heavy oil and bitumen are awarded. In fact, the inaccuracies (i.e., the limits of *experimental difference*) of the method of measuring viscosity also increase the potential for misclassification using this (or any) single property for classification purposes (Smalley, 2000; Speight, 2007, 2009).

In order to classify petroleum, heavy oil, and bitumen, use of a single parameter such as viscosity is inadequate and any attempt to classify petroleum, heavy oil, and bitumen on the basis of a single property is no longer sufficient to define the nature and properties of petroleum and petroleum-related materials.

The generic term *heavy oil* is often applied (for convenience rather than for technical accuracy) to petroleum that has an API gravity of less than 20°, and those highly viscous materials having an API gravity of less than 10° API then have been referred to as *bitumen*. As already stated, use of such a system that relies on one physical property or parameter leads to confusion, and, in fact, the inaccuracies (i.e., the limits of *experimental difference*) of the method of measuring viscosity also increase the potential for misclassification using this (or any) single property for classification purposes.

It is necessary to emphasize then that the tar general classification of petroleum into conventional petroleum, heavy oil, and extra heavy oil involves not only an inspection of several properties but also some acknowledgment of the method of recovery.

In addition, petroleum is generically referred to as a *fossil energy resource* and is further classified as a *hydrocarbon resource* and for illustrative (or comparative) purposes in this text, coal and oil shale kerogen have also been included in this classification. However, the inclusion of coal and oil shale under the broad classification of *hydrocarbon resources* is incorrect insofar as the term *hydrocarbon* be expanded to include the organic high molecular weight (macromolecular) and organic nonhydrocarbon (heteroatomic) species that constitute coal and oil shale. Heteroatomic species are those organic constituents that contain atoms other than carbon and hydrogen, for example, nitrogen, oxygen, sulfur, and metals (nickel and vanadium) as an integral part of the molecular matrix.

Use of the term organic sediments is more correct and to be preferred (Figure 1.2). The inclusion of coal and oil shale kerogen in the category *hydrocarbon resources* is due to the fact that these two natural resources (coal and oil shale kerogen) will produce hydrocarbons on high-temperature processing (Figure 1.3). Therefore, if either coal or oil shale kerogen is to be included in the term *hydrocarbon resources*, it is more appropriate that they be classed as *hydrocarbon-producing resources* under the general classification of *organic sediments*. Thus, fossil energy resources divide into two classes: (1) naturally occurring hydrocarbons (petroleum, natural gas, and natural waxes) and (2) hydrocarbon sources (oil shale and coal) which may be made to generate hydrocarbons by the application of conversion processes. Both classes may very aptly be described as organic sediments.

Whenever attempting to define or classify oil sand bitumen, it is always necessary to return to the definition as given by the United States Federal Energy Administration (FEA) (FE-76-4), presented above.

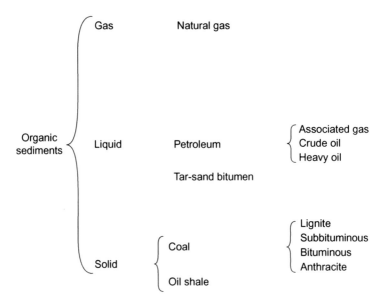

Fig. 1.2 Classification of fossil fuel as organic sediments (Speight, 2007).

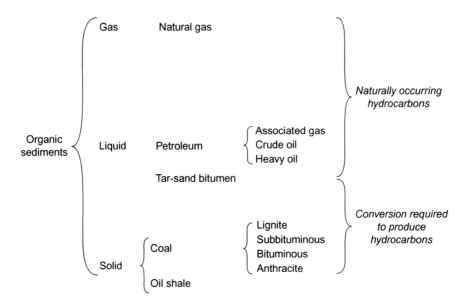

Fig. 1.3 Classification of fossil fuels as hydrocarbon resources and hydrocarbon-producing resources (Speight, 2007).

Thus, by this definition (FE-76-4), oil sand bitumen is not crude oil and it is set apart from conventional crude oil and heavy crude oil insofar as it cannot be recovered from a deposit by the use of conventional (including EOR) oil recovery techniques as set forth in the June 1979 Federal Energy Regulations. To emphasize this point, bitumen has been recovered commercially by mining and the hot water process, and is currently upgraded (converted to synthetic crude oil) by a combination of a thermal or hydrothermal process followed by product hydrotreating to produce a low sulfur hydrocarbon product known as *synthetic crude oil.*

Oil sand bitumen is a naturally occurring material that is immobile in the deposit and cannot be recovered by the application of EOR technologies, including steam-based technologies. On the other hand, heavy oil is mobile in the reservoir and can be recovered by the application of enhanced oil recovery (EOR) technologies, including steam-based technologies.

Since the most significant property of oil sand bitumen is its *immobility* under the conditions of temperature and pressure in the deposit, the interrelated properties of API gravity (ASTM D287) and viscosity (ASTM D445) may present an *indication* (but only an indication) of the mobility of oil or immobility of bitumen, but in reality these properties only offer subjective descriptions of the oil in the reservoir. The most pertinent and objective representation of this oil or bitumen mobility is the *pour point* (ASTM D97) (Chapter 2).

By definition, the *pour point* is the lowest temperature at which oil will move, pour, or flow when it is chilled without disturbance under definite conditions (ASTM D97). In fact, the pour point of an oil when used in conjunction with the reservoir temperature gives a better indication of the condition of the oil in the reservoir than does the viscosity. Thus, the pour point and reservoir temperature present a more accurate assessment of the condition of the oil in the reservoir, being an indicator of the mobility of the oil in the reservoir. Indeed, when used in conjunction with reservoir temperature, the pour point gives an indication of the liquidity of the heavy oil or bitumen and, therefore, the ability of the heavy oil or bitumen to flow under reservoir conditions. In summary, the pour point is an important consideration because, for efficient production, additional energy must be supplied to the reservoir by a thermal process to increase the reservoir temperature beyond the pour point.

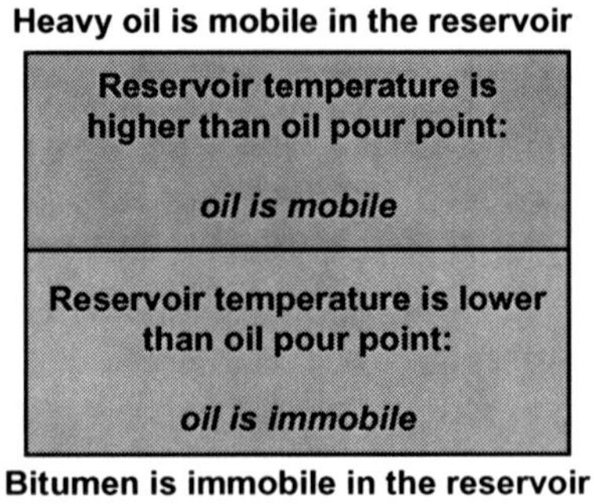

Heavy oil is mobile in the reservoir

Reservoir temperature is
higher than oil pour point:

oil is mobile

Reservoir temperature is lower
than oil pour point:

oil is immobile

Bitumen is immobile in the reservoir

Fig. 1.4 Simplified illustration of the use of pour point and reservoir temperature to differentiate between heavy oil and bitumen (Speight, 2009).

For example, Athabasca bitumen with a pour point in the range 50–100°C (122–212°F) and a deposit temperature of 4–10°C (39–50°F) is a solid or near solid in the deposit and will exhibit little or no mobility under deposit conditions. Pour points of 35–60°C (95–140°F) have been recorded for the bitumen in Utah with formation temperatures on the order of 10°C (50°F). This indicates that the bitumen is solid within the deposit and therefore immobile. The injection of steam to raise and maintain the reservoir temperature above the pour point of the bitumen and to enhance bitumen mobility is difficult, in some cases almost impossible. Conversely, when the reservoir temperature exceeds the pour point, the oil is fluid in the reservoir and therefore mobile. The injection of steam to raise and maintain the reservoir temperature above the pour point of the bitumen and to enhance bitumen mobility is possible and oil recovery can be achieved.

A method that uses the pour point of the oil and the reservoir temperature (Figure 1.4) adds a specific qualification to the term *extremely viscous* as it occurs in the definition of oil sand. In fact, when used in conjunction with the recovery method (Figure 1.5), pour point offers more general applicability to the conditions of the oil in the reservoir or the bitumen in the deposit and comparison of the two temperatures (pour point and reservoir temperatures) shows promise and warrants further consideration.

In summary, heavy oil is more viscous than conventional petroleum but the resources are in plentiful supply and different methods of production are required. However, heavy oil (like conventional petroleum and oil sand bitumen) cannot be defined using a single property. Oil

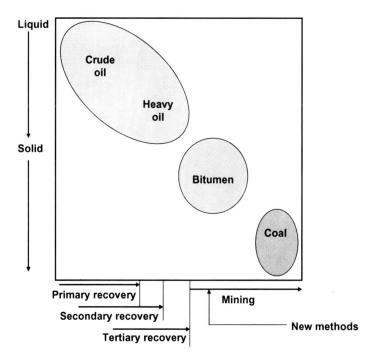

Fig. 1.5 Schematic representation of the properties and recovery methods for crude oil, heavy oil, tar-sand bitumen, and coal (Speight, 2009).

sand bitumen cannot be defined using a single property, but it can be defined by the recovery method; bitumen can be defined by the recovery method. Furthermore, heavy oil is usually mobile in the reservoir, whereas oil sand bitumen is immobile in the deposit.

Large amounts of oil sand bitumen are present globally, even when a conservative viewpoint is taken, but it is not known with precision the proportion of this material ultimately to become recoverable. However, the size of the resource is not a deterrent to exploitation, utilization, or investment in research on production and upgrading.

The workable definitions point to some practical problems in discussing oil sand bitumen because any of these hydrocarbonaceous materials may occur as either surficial deposits, to be strip mined, or subsurface deposits, either to be mined or, more commonly, produced through bore holes with the assistance of added energy.

It is essential to realize that (in the current context) of conventional petroleum and heavy oil, there are several parameters that can

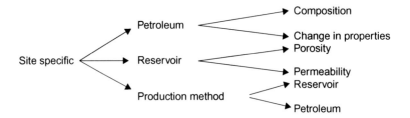

Fig. 1.6 Representation of the changing reservoir and deposit parameters for crude oil, heavy oil, and tar-sand bitumen (Speight, 2007).

influence properties and recovery. These properties are usually site specific to the particular reservoir in which the crude oil or heavy oil is located (Figure 1.6).

Finally, the two conditions of vital concern for the economic development of tar-sand deposits are the concentration of the resource, or the percent bitumen saturation, and its accessibility, usually measured by the overburden thickness. Recovery methods are based on either mining combined with some further processing (Chapter 3) or operation on the oil sands *in situ* (Chapters 4 and 5). The mining methods are applicable to shallow deposits, characterized by an overburden ratio (i.e., overburden depth to thickness of tar-sand deposit) of approximately 1.0. Since the Athabasca oil sands have a maximum thickness of approximately 300 ft, no more than 10% of the in-place deposit is minable within current concepts of the economics and technology of open-pit mining. This 10% portion may be considered as the proved tar-sand reserves.

REFERENCES

Abraham, H., 1945. Asphalts and Allied Substances. Van Nostrand, New York, NY.

Ancheyta, J., Speight, J.G., 2007. Hydroprocessing of Heavy Oils and Residua. CRC, Taylor and Francis Group, Boca Raton, FL.

ASTM D4. Standard Test Method for Bitumen Content. Annual Book of Standards. American Society for Testing and Materials, West Conshohocken, PA.

ASTM D97. Standard Test Method for Pour Point of Petroleum Products. Annual Book of Standards. American Society for Testing and Materials, West Conshohocken, PA.

ASTM D287. Standard Test Method for API Gravity of Crude Petroleum and Petroleum Products (Hydrometer Method). Annual Book of Standards. American Society for Testing and Materials, West Conshohocken, PA.

ASTM D445. Standard Test Method for Kinematic Viscosity of Transparent and Opaque Liquids (and Calculation of Dynamic Viscosity). Annual Book of Standards. American Society for Testing and Materials, West Conshohocken, PA.

ASTM D4175. Standard Terminology Relating to Petroleum, Petroleum Products, and Lubricants. Annual Book of Standards. American Society for Testing and Materials, West Conshohocken, PA.

Attanasi, E.D., Meyer, R.F., 2007. Natural bitumen and extra-heavy oil. In: Trinnaman, J., Clarke, A. (Eds.), 2007 Survey of Energy Resources World Energy Council, London, United Kingdom, pp. 119–143.

Babies, G., Messner, J., 2012. Unconventional Oil. Polinares Working Paper No. 23. Research and Innovation Services, University of Dundee, Dundee, Scotland.

IEA, 2005. Resources to Reserves: Oil & Gas Technologies for the Energy Markets of the Future. International Energy Agency, Paris, France.

Lee, S., 1996. Alternative Fuels. CRC, Taylor and Francis Group, Boca Raton, FL.

Lee, S., Speight, J.G., Loyalka, S., 2007. Handbook of alternative fuel technologies. CRC-Taylor & Francis Group, Boca Raton, Florida.

Meyer, R.F., Attanasi, E.D., Freeman, P.A., 2007. Heavy Oil and Natural Bitumen Resources in Geological Basins of the World. Open File-Report 2007-1084. United States Geological Survey, Reston, VA.

Smalley, C., 2000. Heavy oil and viscous oil. In: Dawe, R.A. (Ed.), Modern Petroleum Technology. John Wiley & Sons, Hoboken, NJ. (Chapter 11).

Speight, J.G., 1984. Upgrading Heavy Oils and Residua: The Nature of the Problem. In: Kaliaguine, S., Mahay, A. (Eds.), Characterization of Heavy Crude Oils and Petroleum Residues. Elsevier, Amsterdam, p. 515.

Speight, J.G., 2000. The Desulfurization of Heavy Oils and Residua, second ed. Marcel Dekker, New York, NY.

Speight, J.G., 2007. The Chemistry and Technology of Petroleum, fourth ed. CRC, Taylor and Francis Group, Boca Raton, FL.

Speight, J.G., 2008. Synthetic Fuels Handbook: Properties, Processes, and Performance. McGraw-Hill, New York, NY.

Speight, J.G., 2009. Enhanced Recovery Methods for Heavy Oil and Tar Sands. Gulf Publishing Company, Houston, TX.

Speight, J.G., 2013. The Chemistry and Technology of Coal, third ed. CRC, Taylor and Francis Group, Boca Raton, FL.

Speight, J.G., Ozum, B., 2002. Petroleum Refining Processes. Marcel Dekker, New York, NY.

CHAPTER 2

Occurrence and Properties

2.1 INTRODUCTION

Oil sand bitumen is a source of liquid fuels that is distinctly separate from conventional petroleum (Attanasi and Meyer, 2007; Meyer and Duford, 1988; Meyer and Schenk, 1985; Meyer et al., 2007; Speight, 1979, 1990, 2005a, 2008, 2009; US Congress, 1976; US DOE, 2007).

Oil sand (also called *tar sand* in many other countries, including the United States) or the more geologically correct term *bituminous sand* is commonly used to describe a sandstone deposit that is impregnated with a heavy, viscous bituminous material. Oil sand is actually a mixture of sand, water, and bitumen, but many of the oil sand deposits in countries other than Canada lack the water layer that is believed to facilitate the hot water recovery process. The heavy bituminous material has a high viscosity under deposit conditions and cannot be retrieved through a well by conventional production techniques.

Geologically, the term *oil sand* is commonly used to describe a sandstone deposit that is impregnated with bitumen, a naturally occurring material that is solid or near solid and is substantially immobile under deposit conditions. The bitumen cannot be retrieved through a well by conventional production techniques, including currently used enhanced recovery techniques (Chapter 1). In addition to this definition, there are several tests that must be carried out to determine whether or not, in the first instance, a resource is an oil sand deposit (Speight, 2001). Most of all, a core taken from an oil sand deposit, and the bitumen isolated therefrom, are certainly not identifiable by the preliminary inspections (sight and touch) alone. However, in the United States, the final determinant is whether or not the material contained therein can be recovered by primary, secondary, or tertiary (enhanced) recovery methods (US Congress, 1976).

The relevant position of oil sand bitumen in nature is best illustrated by comparing its position relative to petroleum and heavy oil. Thus, petroleum is generically referred to as a *fossil energy resource* (Chapter 1) and is further classified as a hydrocarbon resource and, for illustrative (or comparative) purposes in this report, coal and oil shale kerogen are also included in this classification. However, the inclusion of coal and oil shale under the broad classification of hydrocarbon resources has required (incorrectly) that the term hydrocarbon be expanded to include these resources. It is essential to recognize that resources such as coal, oil shale kerogen, and oil sand bitumen contain large proportions of heteroatomic species. Heteroatomic species are those organic constituents that contain atoms other than carbon and hydrogen, for example, nitrogen, oxygen, sulfur, and metals (nickel and vanadium).

Detailed characterization of all of the above aspects is required to better understand the interrelationship of components within the deposit and the overall deposit behavior and reactivity at production conditions, for optimization of exploitation methods and the development of new technologies. Studies should focus on the heterogeneous distribution and composition of minerals and the composition of the bitumen.

2.2 OCCURRENCE AND RESERVES

Oil sand deposits are widely distributed throughout the world in a variety of countries (Attanasi and Meyer, 2007; Gutierrez, 1981; Meyer and Dietzman, 1981; Meyer et al., 1984; Phizackerley and Scott, 1978).

The various oil sand deposits have been described as belonging to two types: (1) materials that are found in *stratigraphic* traps and (2) deposits that are located in *structural* traps. There are inevitably gradations and combinations of these two types of deposits and a broad pattern of deposit entrapment are believed to exist.

The distinction between a *structural trap* (the usually description of a petroleum reservoir) and a *stratigraphic trap* is often not clear. For example, an *anticlinal trap* may be related to an underlying buried limestone reef. Beds of sandstone may wedge out against an anticline because of depositional variations or intermittent erosion intervals. Salt domes, formed by flow of salt at substantial depths, also have created numerous traps that are both a structural trap and a stratigraphic trap.

In general terms, the entrapment characteristics for the very large oil sand deposits all involve a combination of stratigraphic and structural traps. Entrapment characteristics for the very large oil sands all involve a combination of stratigraphic–structural traps. There are no very large oil sand accumulations having more than four billion barrels (4×10^9 bbl) in place either in purely structural or in purely stratigraphic traps. In a regional sense, structure is an important aspect since all of the very large deposits occur on gently sloping homoclines.

Furthermore, the oil sand deposits of the world have been described as belonging to two types. These are: (1) *in situ* deposits resulting from breaching and exposure of an existing geologic trap and (2) migrated deposits resulting from accumulation of migrating material at outcrop (Speight, 2005b). However, there are inevitable gradations and combinations of these two types of deposits. The deposits were laid down over a variety of geologic periods and in different entrapments and a broad pattern of deposit entrapment is believed to exist since all deposits occur along the rim of major sedimentary basins and near the edge of pre-Cambrian shields. The deposits either transgress an ancient relief at the edge of the shield (e.g., those in Canada) or lie directly on the ancient basement (e.g., as in Venezuela, West Africa, and Madagascar).

A feature of major significance in at least five of the major areas is the presence of a regional cap (usually widespread transgressive marine shale). Formations of this type occur in the Colorado Group in Western Canada, in the Freites formation in eastern Venezuela, or in the Jurassic formation in Melville Island overlying the formation. The cap plays an essential role in restraining vertical fluid escape from the basin thereby forcing any fluids laterally into the paleo-delta itself. Thus, the subsurface fluids were channeled into narrow outlets at the edge of the basin.

Sedimentary structures observed in the Athabasca oil sands include some large- and very large-scale features, such as deltaic foreset bedding, planar and trough cross-stratification, and many small-scale features including micro-cross-lamination and structures due to disruption and disturbance of horizontal laminae of silt by slumping, load casting, faulting, and burrowing. These features are believed to have been formed in environments of deposition which ranged from fluviatile at the base, through shallow water deltaic in the middle, to open-sea marine at the top. This sequence of strata is thought to have been laid down in quick succession and preserved without major erosion. Predominance of lateral processes in the accumulation of the strata in this deposit makes local bed-to-bed correlation difficult and knowledge of the primary sedimentary structures is of practical importance in the economic exploration and exploitation of this vast petroleum deposit (Carrigy, 1967; Flach, 1984).

The potential reserves of hydrocarbon liquids (available through conversion of the bitumen to *synthetic crude oil*) that occur in oil sand deposits have been variously estimated on a world basis to be in excess of three trillion (3×10^{12}) barrels. However, the issue is whether or not these reserves can be recovered and converted to synthetic crude oil. Geographical and geological feature may well put many of these resources beyond the capabilities of current recovery technology, requiring new approaches to recovery and conversion.

2.2.1 Canada

The oil sand resources of Canada are situated almost entirely within the province of Alberta (Figure 2.1), with only minor oil sand deposits found on Melville Island in the Arctic Island region of Canada. These deposits lie on the north shore of Marie Bay, Melville Island,

Fig. 2.1 Location of tar-sand deposits in Alberta, Canada. www.ags.gov.ab.ca.

some 1,450 miles north of Edmonton; Triassic sandstone of the Bjorn Formation are impregnated with a bituminous material. This deposit was discovered in 1962 and the sands occur at intervals along a 60 mile outcrop. The richer sands tend to be associated with structurally high areas or are closely related with faults, and with minor occurrences of oil shale on the eastern edge of the Western Canada Sedimentary Basin (ERCB, 2011).

There are four major reserves of oil sands in the province—Peace River, Athabasca, Wabasca (often spelled Wabiskaw because of the Wabiskaw sandstone formation), and Cold Lake (Speight, 2005b).

Together, they underlie nearly 30,000 square miles. The oil sands occur in layers at different depths in each deposit and all of the oil sand deposits are composed mainly of sand (quartz, mica, rutile, zirconia, and pyrite) along with clay, water, and bitumen. More than 10% bitumen is considered rich oil sand, from 6% to 10% moderate, and less than 6% lean. The Alberta oil sands are *hydrophilic* (water wet)—each grain of sand is surrounded by an envelope of water which, in turn, is surrounded by bitumen and it is this feature which makes the hot water separation process practical.

In the *Peace River deposit*, the crude bitumen occurs at a depth of from 1,000 to 2,500 ft in the Bluesky and Gething formation. In an area of 3,000 square miles, there are an estimated 71.7 billion barrels (71.7×10^9 bbl) of bitumen in place.

The Wabasca oil sand deposit is usually indicated on maps as part of the Athabasca reserve, but they are not connected. The Wabasca oil sand is at a higher stratigraphic level, but under a greater depth of overburden than the mineable portion of the Athabasca reserves. The layers of reserves are thinner and more deeply buried, with an average thickness of about 7 ft lying under 490–1,500 ft of overburden. Underlying an area of 2,600 square miles, the deposit contains an estimated 13% oil saturation and holds in place 42.5 billion barrels (42.5×10^9 bbl) of bitumen. Because of the depth, none of this is available to existing surface mining methods.

The Athabasca oil sands are the largest and most accessible reserve of bitumen; the sands underlie an area of 16,000 square miles and contain an estimated 812 billion barrels (812×10^9 bbl) of bitumen in place in the McMurray formation. The deposit has three layers of oil-bearing sand, one above the other, separated by beds of silt, sand, and shale. The bitumen accumulation is rich and thick, and covered by overburden up to 2,500 ft deep, consisting of muskeg, glacial tills, sandstone, and shale. Approximately 7% of the deposit lies under less than 250 ft of overburden, making it accessible to surface mining techniques. This deposit contains an estimated 135 billion barrels (135×10^9 bbl) of the bitumen in place. An estimated 33 billion barrels (33×10^9 bbl) of bitumen reserves can be recovered by present mining methods.

The Cold Lake oil sands consist of four separate reservoirs/deposits—one each in the McMurray, Clearwater, Lower Grand Rapids, and upper Grand Rapids formations. The bitumen is contained within thick, loosely consolidated sandstone beds. This, along with the depth, which varies from 984 to 1,969 ft, prohibits surface mining but makes deposits suitable for *in situ* extraction methods. The crude bitumen in these deposits is the least viscous of any oil sand but still denser than the heavy oils located to the south. Approximately 8,200 square miles in area, the Cold Lake sands hold an estimated 178 billion barrels (178×10^9 bbl) of bitumen in place and potentially can yield 28 billion barrels (28×10^9 bbl) of synthetic crude oil.

The Alberta government has estimated that the Athabasca deposit along with the neighboring Wabasca and Peace River deposits contain approximately two trillion barrels (2×10^{12} bbl) of bitumen. Of this total, approximately 174 billion barrels (174×10^9 bbl) of crude bitumen are economically recoverable from the three Alberta oil sands areas at current prices using current technology (ERCB, 2011; Speight, 2005b). The estimates of the bitumen reserves (and the possible synthetic crude oil) place Canadian proven oil reserves second in the world behind those of Saudi Arabia.

According to the acceptable definition in use by the government of the United States, the final determinant of whether or not a material is oil sand depends upon the recovery method required (Chapter 1). In fact, the bitumen in the Athabasca (Alberta, Canada) deposit fits the definition very well insofar as the material is recovered by a mining operation and not by any of the recognized enhanced recovery techniques.

The key characteristic of the Athabasca deposit is that it is the only one shallow enough to be suitable for surface mining (Chapter 3). Approximately 10% of the Athabasca oil sands are covered by less than 250 ft of overburden. The mineable area as defined by the Alberta government covers 37 contiguous townships (approximately 1,300 square miles) north of the city of Fort McMurray. The overburden consists of 3–10 ft of waterlogged muskeg on top of 0–250 ft of clay and barren sand, while the underlying oil sands are typically 130–200 ft thick and sit on top of relatively flat limestone rock. As a result of the easy accessibility, the world's first oil sands mine was started by Great Canadian Oil Sands (GCOS) Limited (a predecessor company of Suncor Energy) in 1967. The Syncrude mine (the biggest mine in the world) followed in 1978, and the Albian Sands mine (operated by Shell Canada) in 2003.

All three of these mines are associated with bitumen upgraders that convert the unusable bitumen into synthetic crude oil for shipment to refineries in Canada and the United States, though in the case of the Albian project, the upgrader is not colocated with the mine, but at Scotford, 275 miles south of the mine site. The bitumen, diluted with a solvent, is transferred there from the mine site to the upgrader by pipeline.

In the context of the Athabasca deposit, inconsistencies arise presumably because of the lack of mobility of the bitumen at formation temperature (approximately 4°C, 39°F). For example, the proportion of bitumen in the oil sand increases with depth within the formation. Furthermore, the proportion of the nonvolatile asphaltenes or the nonvolatile asphaltic fraction (asphaltenes plus resins) in the bitumen also increases with depth within the formation, leading to reduced yields of distillate from the bitumen obtained from deeper parts of the formation. In keeping with the concept of higher proportions of asphaltic fraction (asphaltene constituents plus resin constituents), variations (horizontal and vertical) in bitumen properties have been noted previously, as have variations in sulfur content, nitrogen content, and metals content (Speight, 1999).

Obviously, the richer oil sand deposits occur toward the base of the formation. However, the bitumen is generally of poorer quality than the bitumen obtained from near the top of the deposit insofar as the proportions of nonvolatile coke-forming constituents (asphaltene constituents plus resin constituents) are higher (with increased proportions of nitrogen, sulfur, and metals) near the base of the formation.

2.2.2 United States

In the United States, there are several deposits of bitumen that are characterized as giant deposits which, for convenience, may be considered to be those containing at least one billion barrels (1.0×10^9 bbl) of bitumen. Examples of such deposits are (Meyer and de Witt, 1990):

1. Arctic Coastal Plain basin, Alaska, Kuparuk deposit, Paleocene, and Upper Cretaceous Ugnu sandstones (Werner, 1987).
2. Cherokee basin, Kansas and Missouri, sandstones of the Middle Pennsylvanian Bluejacket, and Warner Sandstone Members of the Krebs Formation.
3. Gulf Coast basin, Texas, San Miguel deposit, and sandstones of the Upper Cretaceous San Miguel Formation.

4. Illinois basin, Kentucky, and Upper Mississippian Big Clifty Sandstone Member of the Golconda Formation.
5. Paradox basin, Utah, Tar Sand Triangle deposit, Lower Permian White Rim Sandstone Member of the Cutler Formation and Circle Cliffs deposit, and sandstones of the Middle Triassic Moenkopi Formation.
6. Santa Maria basin, California, Foxen deposit, sandstones of the Pliocene Foxen Formation and Casmalia field, diatomaceous mudstone of the Miocene, and Pliocene Sisquoc Formation.
7. Uinta basin, Utah, PR Spring deposit, sandstones of the Paleocene and Eocene Green River Formation and Sunnyside deposit, sandstones of the Paleocene, and Eocene Green River Formation.
8. Black Warrior basin, Alabama, and Upper Mississippian Hartselle Sandstone.
9. Carter and Murray Counties, South-Central Oklahoma (Harrison and Burchfield, 1983.)

However, the major oil sand deposits of the United States occur within and around the periphery of the Uinta Basin, Utah (Blackett, 1996) (Figure 2.2). These include the Sunnyside, Oil sand Triangle, Peor Springs (PR Springs), Asphalt Ridge, and sundry other deposits (Table 2.1). Asphalt Ridge lies on the north eastern margin of the central part of the Uinta Basin at the contact of the tertiary beds with the underlying Cretaceous Mesaverde Group. The Mesaverde (Mesa Verde) Group is divided into three formations: two of these, the Asphalt Ridge sandstone and the Rim Rock sandstone, are beach deposits containing the viscous bitumen. The Rim Rock sandstone is thick and uniform with good deposit characteristics and may even be suitable for thermal recovery methods. The Duchesne River formation (Lower Oligocene) also contains bituminous material, but the sands tend to be discontinuous.

Asphalt Ridge has been characterized as holding approximately one billion barrels (1×10^9 bbl) of recoverable oil with the potential to support a 50,000 bbl/day facility. However, steady growth of the community of Vernal has encumbered some of the resource. Two rich locations could produce significant yields of bitumen but in more modest quantities than originally contemplated. The technology used for the Alberta sands could be adaptable for use in the unconsolidated sands of the rich zones.

The Sunnyside deposits extend over a greater area than Asphalt Ridge and are located on the southwest flank of the Uinta Basin. The oil sand

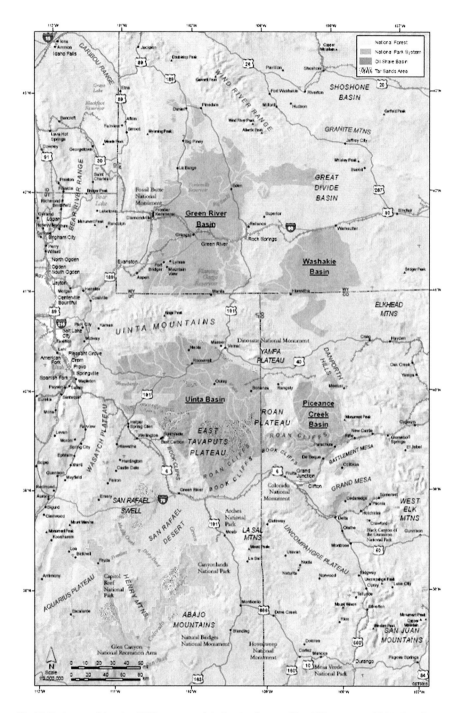

Fig. 2.2 Utah tar-sand deposits. US Department of the Interior, Bureau of Land Management (BLM). http://ostseis.anl.gov/guide/tarsands/index.cfm.

Table 2.1 Estimated Bitumen Reserves of the Utah Tar-Sand Deposits		
Deposit	Known Reserves (bbl×10⁶)	Additional Probable Reserves (bbl×10⁶)
Sunnyside	4,400	1,700
Tar Sand Triangle	2,500	420
PR Spring	2,140	2,230
Asphalt Ridge	820	310
Circle Cliffs	590	1,140
Other	1,410	1,530
Total	118,060	7,330

Source: US Department of the Interior, Bureau of Land Management (BLM). http://ostseis.anl.gov/guide/tarsands/index.cfm.

accumulations occur in sandstone of the Wasatch and lower Green River formations (Eocene). The Wasatch sandstone contains bitumen impregnation but is lenticular and occupies broad channels cut into the underlying shale and limestone; the Green River beds are more uniform and laterally continuous. The source of the bitumen in the Asphalt Ridge and Sunnyside accumulation is considered to be the Eocene Green River shale. The Sunnyside deposit contains enough recoverable resource to support a 100,000 bbl/day operation. Thermal or solvent treatment may be required as the ore is consolidated.

The Peor Springs accumulation is about 60 miles (96.5 km) east of the Sunnyside deposit and occurs as lenticular sandstone (Eocene Wasatch formation). There are two main beds from 30 to 85 ft (9 m to 26 m) thick with an estimated overburden thickness of 0–250 ft (0–76 m). The tilt of the southern flank of the Uinta Basin has left this deposit relatively undisturbed except for erosion, which has stripped it of its cover allowing the more volatile constituents to escape. In the central southeast area of Utah, some deposits of bitumen-impregnated sandstone occur in Jurassic rock, but the greater volume of in-place bitumen occurs in rocks of Triassic and Permian age. This sizeable resource is close to the surface but is fragmented by erosion and multiple beds. It is in a primitive area, which may slow development. A few rich zones could each support modestly sized operations on the order of 25,000–50,000 bbl/day.

The Tar Sand Triangle—located near Canyon Lands National Park—is considered to be a single, giant stratigraphic trap containing the bitumen. The bitumen is characterized by high sulfur content, similar to Alberta oil sands but unlike the Uinta Basin deposits described above, which are low in sulfur. There appears to be interest in this deposit for *in situ* recovery. The product could be transported by truck and rail in the bitumen or diluted bitumen state.

The Californian deposits are concentrated in the coastal region west of the San Andreas Fault (De Chadenedes, 1987; Dibblee et al., 1987; Hallmark, 1982). Numerous deposits are estimated to contain a cumulative total of eight billion barrels (8×10^9 bbl) of bitumen in place.

The largest deposit is the Edna deposit, which consists of conglomerate, sandstone, diatomaceous sandstone, and siliceous shale. The deposit occurs as a stratigraphic trap and outcrops in scattered areas on both flanks of a narrow syncline. The deposit extends over an area of about 7,000 acres and occurs from outcrop to 100 ft (30 m) depth. The accumulations are considered to have been derived from the underlying organic and proliferous Monterey shale.

Generally, recorded estimates have been concerned with strip mining potential and the studies limited to deposits no deeper than about 250 ft. However, the Edna deposit extends to much greater depths where the bitumen is fluid (due to a high deposit temperature) and is produced by wells from depths of 500–1,500 ft. Tests conducted on the shallow deposits yielded 9–16% w/w bitumen. The total reserve, estimated to a depth of 250 ft, was approximately 175 million barrels of bitumen in place. Total resources in place may be many times greater than this estimate, because more recent wells have penetrated as much as 1,200 ft of oil sand.

The Foxen Tar Sand lies at the base of the Foxen Formation (Pliocene; the base is the Foxen-Sisquoc unconformity) and extends throughout a large area of the Santa Maria Valley. The top of the oil sand varies in depth from 500 ft to approximately 4,000 ft, with an average net thickness of approximately 100 ft. The sand is fine grained with an average porosity of 33%, an average bitumen saturation of 30%, and an average permeability of about 300 mD. The bitumen varies from 9° to 14° API. A complete evaluation of this deposit is not available, but the widespread occurrence of this oil sand in wells throughout the area suggests a very significant accumulation and the bitumen in place is speculatively estimated to be on the order of two billion barrels (2×10^9 bbl).

The Sisquoc area (located approximately 20 miles southeast of Santa Maria and south of the Sisquoc River) has several discontinuous outcrops of rich oil sand, which was mined for asphalt as early as 1894. The maximum thickness of the deposit (Careaga Sandstone, late Pliocene) is on the order of 185 ft, with an average thickness of approximately 85 ft. The overburden reaches a maximum thickness of 70 ft, with an average thickness of 15 ft. Within the 175 acres of this deposit are an estimated 29 million barrels of bitumen, although other estimates reach as high as one million barrels of bitumen. Another bituminous rock deposit, estimated to contain more than 56 million barrels of bitumen, is located on the northern side of the Sisquoc River in La Brea Creek Valley.

The Sisquoc deposit (Upper Pliocene) is the second largest in California and occurs in sandstone in which there are as many as eight individual oil sand units. The total thickness of the deposit is about 185 ft (56 m) occurring over an area of about 175 acres with an overburden thickness between 15 and 70 ft. The sands lie above the Monterey shale, which has been suggested to be the source of the bitumen.

The third California deposit at Santa Cruz is located approximately 56 miles (90 km) from San Francisco. The material occurs in sandstone of the Monterey and Vaqueros formations, which are older than both the Edna and Sisquoc deposits. The Santa Cruz oil sands are discontinuous and overlie the pre-Cretaceous basement.

The Oxnard oil field is a large and currently productive oil field in and adjacent to the city of Oxnard, Ventura County, California. While the conventional oil reservoir is seriously depleted, there is a large oil sand deposit amounting to a potential 400 million barrels of oil equivalent (Hallmark, 1982).

The Oxnard field, discovered in January 1937 by Vaca Oil Exploration Company, is part of a structural down warp that occurred during the late Pliocene with rocks that are all sedimentary and mostly marine. Oil accumulations, of which there are many in the Ventura province, mainly occur in anticlinal settings modified by faulting, but stratigraphy is also influential in creating traps in the area.

Under the surface alluvium, a series of relatively impermeable sedimentary units cap the petroleum-bearing formation—on top are the Pleistocene, San Pedro, and Pleistocene-upper Pliocene Santa Barbara formations under which the Pliocene Pico Sands contain areas of oil sands that, separated by an unconformity (the Miocene Monterey

Formation), also contains an oil sand deposit (Vaca Tar Sand) (CDC, 1998; Hallmark, 1982; Keller, 1988).

The Vaca Tar Sand is found at depths from 1,800 to 2,500 ft, and the rock characteristics are typical of most California reservoirs and deposits (high temperature relative to the heavy oil or bitumen), making the reservoirs/deposits excellent candidates for thermal enhanced recovery methods. However, the extremely poor quality of this crude oil (5–8° API) and high sulfur content (7.5% w/w) have held up development.

South Texas holds the largest oil sand deposits. These deposits occur in the San Miguel tar belt (Upper Cretaceous) mostly in Maverick and Zavala counties as well as in the Anacadro limestone (Upper Cretaceous) of the Uvalde district.

The Kentucky oil sand deposits are located at Asphalt, Davis-Dismal Creek, and Kyrock; they all occur in nonmarine Pennsylvanian or Mississippian sediments. The three deposits appear as stratigraphic traps and are thought to have received their bitumen or bitumen precursor from the Devonian Chattanooga shale. Oil sand deposits in New Mexico occur in the Triassic Santa Rosa sandstone, which is irregularly bedded, fine- to medium-grained micaceous sandstone (Noger, 1999).

Finally, in the context of the oil sand deposits in the United States, those in Missouri occur over an area estimated at 2,000 square miles in Barton, Vernon, and Cass Counties and the sandstone bodies that contain the bitumen are middle Pennsylvanian in age. The individual bitumen-bearing sands are approximately 50 ft (15 m) in thickness except where they occur in channels which may actually be as much as 250 ft (76 m) thick. The two major deposits are the Warner sandstone and the Bluejacket sandstone that at one time were regarded as blanket sands covering large areas. However, recent investigations suggest that these sands can abruptly grade into barren shale or siltstone.

Oil sands in the various areas of the United States such as Alaska, Alabama, Texas, California, and Kentucky are deeper and/or thinner than the Utah deposits and less economic to develop.

2.2.3 Venezuela

Oil sand deposits in Venezuela occur in the Officina-Tremblador tar belt, which is believed to contain bitumen-impregnated sands of a similar extent to those of Alberta, Canada. The Officina formation overlaps the Tremblador (Cretaceous) formation and the organic material is typical bitumen having API gravity less than 10°. The Guanaco Asphalt Lake occurs in deposits that rest on a formation of mid-Pliocene age. This formation, the Las Piedras, is principally brackish sandstone to freshwater sandstone with associated lignite. The Las Piedras formation overlies a marine Upper Cretaceous group; the Guanaco Lake asphalt is closely associated with the Guanaco crude oil field that produces heavy crude oil from shale and fractured argillite of the Upper Cretaceous group.

The geological setting of the Orinoco deposit is very complex, having evolved through three cycles of sedimentation. Both structural and stratigraphic traps depend on the location, the age of the sediment, and the degree of faulting contain the bitumen. The oil sands are located along the southern flanks of the eastern Venezuelan basin where three distinct zones are apparent from north to south: a zone of tertiary sedimentation, a central platform with transgressive overlapping sediments, and a zone of Proterozoic erosion remnants covered by sediments. The deposit also contains three systems of faulting. All the faults are normal and many are concurrent with deposition (Speight, 2005b).

The Orinoco oil sands are known to be one of the largest deposits, second only to the Athabasca deposits. The Orinoco deposits have been variously estimated to contain approximately 1,200 billion barrels (1.2×10^9 bbl). Petróleos de Venezuela has estimated that the producible reserves of the Orinoco Belt are up to 235 billion barrels (235×10^9 bbl). The United States Geological Survey (USGS) has updated this estimate to 513 billion barrels (513×10^9 bbl) of bitumen, which is recoverable (producible using currently available technology and industry practices) (USGS, 2009).

The Venezuelan Orinoco oil sands site may contain more material than the Athabasca sands. However, while the Orinoco deposits are less viscous and more easily exploited using conventional techniques (the Venezuelan government prefers to call them extra heavy oil), they are too deep to access by surface mining.

2.2.4 Other Areas of Latin America

In Mexico, there are four zones of accumulation of heavy crudes, all of them located along the Gulf Coast (Gutierrez, 1981; Speight, 2005b).

The accumulations are in the areas of Tampico, Chiapas, and Tasco. Some of them have been partially exploited. Although data on the size of these resources remains minimal, on the basis of the volumes of sediments in the basins and the published global figures on probable reserves, it is reasonable to estimate those resources in terms of thousands of millions of barrels of oil in place.

In Colombia and Ecuador, accumulations of heavy crudes are postulated in the Meta Basin and in the Napo-Putumayo and Maranon Sub-Basins. In Colombia, the most prospective zones are situated in the so-called Rio L area, at depths of up to 14,000 ft and in the Villavicencio area at depths of up to 7,500 ft, where crudes of 12–14° API are expected to be found.

With regard to Ecuador, records indicate important accumulations of heavy crudes in the 12–18° API range.

Peru possesses large basins within the structural complex lying between the Guiana and Brazilian Shields, on the east, and the nucleus formed by the Andean Ranges, on the west. Between these limits are aligned five sedimentary systems, with thicknesses greater than 10,000 ft. They are the Marafion, Lower Ucayali, and Upper Ucayali Sub-Basins, and the Beni-Madre de Dios and Tarija-Santa Cruz Basins. In Peru, the best studied basin is that of Marafion, where 17 structures have been tested, of which three, in the Cretaceous Vivian Formation, produced heavy crude oil having 15–17° API. In the same basin, toward the South, the Bretana area was tested at a depth of 2,000 ft and produced 800 bbl/day of 17° API heavy oil.

In Chile, the deposits of relative importance are found in the Province of Antofagasta. There is no information on volumes, but it is possible that new studies may reveal appreciable quantities of this resource.

With regard to Argentina, deposits of petroliferous shales are mentioned for the areas of Mendoza, Malarge, Uspallata, Rio Blanco, and Santa Clara. The volume of shale has been estimated at about 800 million tons, from which could be extracted some 500 million (5×10^8) barrels of oil. In spite of the fact that in the southeast as well as in the northwest of Paraguay parts of sedimentary basins are present, the literature consulted makes no mention of the occurrence of petroliferous deposits in that region. The oil-bearing Permian shales, located in the South of Brazil, extend into the northern part of Uruguay.

Brazil probably possesses the largest number of petroliferous shale accumulations in Latin America. The deposits are to be found distributed in rocks of differing ages, from Paleozoic to Tertiary. They extend along a great arc in the north, west, and south of the country. The largest accumulations are constituted by the Permian Shales in the extreme south. For some of the accumulations, volumes have been cited of the order of 800 billion (800×10^9) tons of oil-bearing shales and a probable recovery of some 50 thousand million (50×10^9) barrels of oil.

There is a small accumulation at Chumpi, near Lima (Peru), which occurs in tuffaceous sands and it is believed to be derived from strongly deformed Cretaceous limestone from which a petroleum type was distilled as a result of volcanic activity.

2.2.5 Other Countries
Oil sand deposits occur throughout the world and older records show that such deposits occur in many countries (Speight, 2005b).

The Bemolanga (Madagascar) deposit is the third largest oil sand deposit presently known and extends over some 150 square miles in western Madagascar with a recorded overburden from 0 to 100 ft (0–30 m). The average pay zone thickness is 100 ft (30 m) with a total bitumen in place quoted at approximately two billion barrels (2×10^9 bbl). The deposit is of Triassic age and the sands are cross-bedded continental sediments; the coarser, porous sands are more richly impregnated. The origin of the deposit is not clear; the most preferred source is the underlying shale or in down-dip formations implying small migration (Speight, 2005b).

The largest oil sand deposit in Europe is that at Selenizza, Albania. This region also contains the Patos oil field throughout which there occurs extensive bitumen impregnation. This deposit occurs in middle-upper Miocene lenticular sands, characterized by a brackish water fauna. Succeeding Pliocene conglomerate beds, which are more generally marine, are also locally impregnated with heavy crude oil. The Selenizza and Patos fields occupy the crestal portions of a north-south trending anticline. Faulting also controls the vertical distribution of the accumulation. The Miocene rests on Eocene limestone and it is these that are thought by some to be the source of the tar.

The Trinidad Asphalt (Pitch) Lake situated on the Gulf of Paria, 12 miles west southwest of San Fernando and 138 ft (43 m) above sea level, occupies a depression in the Miocene sheet sandstone. It overlies an eroded anticline of Upper Cretaceous age with remnants of an early tertiary formation still preserved on the flanks.

The Trinidad bitumen is currently mined and sold as road asphalt. Estimates of the amount available vary and a very approximate estimate indicates that at current production rates (ca. 50,000 tons per year) there is sufficient to last 150 years. There are no current plans to use the Trinidad bitumen as a source of synthetic fuels.

The Romanian deposits are located at Derna deposits and occur (along with Tataros and other deposits) in a triangular section east and northeast of Oradia between the Sebos Koros and Berrettyo rivers. The oil sand occurs in the upper part of the Pliocene formation and the asphalt is characterized by its penetrating odor. The rock is nonmarine, representing freshwater deposition during a period of regression.

Oil sand deposits occur at Cheildag, Kobystan, and outcrop in the south flank of the Cheildag anticline; there are approximately 24 million barrels (24×10^6 bbl) of bitumen in place. Other deposits in the former USSR occur in the Olenek anticline (northeast of Siberia) and it has been claimed that the extent of asphalt impregnation in the Permian sandstone is of the same order of magnitude (in area and volume) as that of the Athabasca deposits. Oil sands have also been reported from sands at Subovka, and the Notanebi deposit (Miocene sandstone) is reputed to contain 20% bitumen by weight. On the other hand, the Kazakhstan occurrence, near the Shubar-Kuduk oil field, is a bituminous lake with a bitumen content that has been estimated to be of the order of 95% by weight of the deposit.

Oil sand occurrences also occur in the Southern Llanos of Colombia where drilling has presented indications of deposits generally described as heavy crude oil, natural asphalt, and bitumen. Most of these occurrences are recorded below 1,500 ft. The oil sands at Burgan in Kuwait and at the Inciarte and Bolivar coastal fields of the Maracaibo Basin are of unknown dimensions. Those at Inciarte have been exploited and all are directly or closely associated with large oil fields. The oil sands of the Bolivar coastal fields are above the oil zones in Miocene beds and are in a lithological environment similar to that of the Officina-Tremblador tar belt. The small Miocene asphalt deposits in the Leyte Islands (Philippines) are extreme samples of stratigraphic entrapment and resemble some of the Californian deposits. Those of the Mefang Basin in Thailand are in Pliocene beds that overlie Triassic deposits and their distribution is stratigraphically controlled.

Finally, oil sand deposits have also been recorded in Spain, Portugal, Cuba, Argentina, Thailand, and Senegal, but most are poorly defined and are considered to contain (in place) less than 1 million barrels (1×10^6 bbl) of bitumen.

2.3 MINERALOGY AND PROPERTIES OF OIL SAND

Oil sand is a mixture of sand and other rock materials that is composed of approximately 80–85% sand, clay (aluminosilicate minerals), and other mineral matter, 5–10 wt% water, and 1–18 wt% bitumen. The use of the term clay, as used in this text, is based on a size classification and is usually determined by a sedimentation method.

However, as might be expected from these statements, oil sand deposits are not uniform. Differences in the permeability and porosity of the deposit and varying degrees of alteration by contact with air, bacteria, and groundwater mean that there is a large degree of uncertainty in the estimates of the bitumen content of a given oil sand deposit.

The oil sand deposits (particularly the Athabasca deposit) are composed primarily of quartz sand, silt and clay, water, and bitumen, along with minor amounts of other minerals, including titanium, zirconium, tourmaline, and pyrite. Although there can be considerable variation, a typical composition is (1) 75–80% w/w inorganic material, with this inorganic portion comprised of 90% w/w quartz sand, (2) 3–5% w/w water, and (3) 1–18% w/w bitumen, with bitumen saturation varying between 0% and 18% by weight. The oil sands are generally unconsolidated and thus quite friable and crumble easily in the hand.

Bitumen is a thick, viscous carbonaceous material that, at room temperature, is in a near-solid state and which is immobile in the deposit. In addition to high viscosity, the bitumen contained in the oil sands is characterized by high densities, high concentration of metals, and a high ratio of carbon-to-hydrogen molecules in comparison with conventional crude oil. With a density range of 0.970–1.015 (8–14° API), and a viscosity at room temperature typically greater than 50,000 cP, bitumen is a thick, black, tar-like substance that pours extremely slowly.

Bitumen is deficient in hydrogen (and approximately 10% w/w hydrogen), when compared with typical crude oil, which contains approximately 12–14% w/w hydrogen. Therefore, to make it an acceptable feedstock for conventional refineries, it must be upgraded through the addition of hydrogen or the rejection of carbon. In order to transport bitumen to refineries equipped to process it, bitumen must be blended with a diluent, traditionally naphtha, to meet pipeline specifications for density and viscosity. The naphtha must contain high proportions of cyclic constituents and/or aromatic constituents to prevent the asphaltene constituents from separating, as would be the case with naphtha, if the naphtha was composed predominantly of paraffinic constituents.

Prior to processing, the bitumen is separated from the sand, as well as other minerals and formation water before it is delivered to downstream refineries. Shallow oil sands deposits, less than about 250 ft (75 m) to the top of the oil sands zone, are exploited using surface mining to recover ore-grade oil sands, which are then delivered to an extraction plant for separation of bitumen from the sand, other minerals, and connate water. Deep oil sands, greater than about 250 ft (75 m) to the top of the oil sands zone, are exploited using *in situ* recovery techniques, whereby the bitumen is separated from the sand *in situ* and produced to the surface through wells drilled into the oil sands deposit.

2.3.1 Mineralogy

By definition, oil sand is a mixture of sand, water, and bitumen with the sand component occurring predominantly as quartz (Strausz and Lown, 2003). The arrangement of the sand, water, and bitumen has been assumed to be an arrangement whereby each particle of the sand is water wet and a film of bitumen surrounds the water-wetted grains. The balance of the void volume is filled with bitumen, connate water, or gas; fine material, such as clay, occurs within the water envelope.

The mineralogy of oil sand deposits does affect the potential for recovery of the bitumen (Strausz and Lown, 2003). Usually, more than 99% by weight of the tar-sand mineral is composed of quartz sand and clays. In the remaining 1%, more than 30 minerals have been identified, mostly calciferous or iron based. Particle size ranges from large grains (99.9% is finer than 1,000 μm) down to 44 μm (325 mesh), the smallest size that can be determined by dry screening. The size between 44 and 2 μm is referred to as silt; sizes below 2 μm (equivalent spherical diameter) are clay.

The Canadian deposits are largely unconsolidated sands with a porosity ranging up to 45% and have good intrinsic permeability. However, the deposits in the United States in Utah range from predominantly low-porosity, low-permeability consolidated sand to, in a few instances, unconsolidated sands. In addition, the bitumen properties are not conducive to fluid flow under normal deposit conditions in either the Canadian or the United States deposits. Nevertheless, where the general nature of the deposits prohibits the application of a mining technique (as in many of the United States deposits), a nonmining technique may be the only feasible bitumen recovery option.

One additional aspect of the character of Athabasca oil sands is that the sand grains are not uniform in character. Grain-to-grain contact is variable and such a phenomenon influences attempts to repack mined sand, as may be the case in studies involving bitumen removal from the sand in laboratory-type *in situ* studies. This phenomenon also plays a major role in the expansion of the sand during processing where the sand to be returned to the mine site might occupy 120–150% of the volume of the original as-mined material.

The oil sand mass can be considered a four-phase system composed of solid phase (siltstone and clay), liquid phase (from fresh to more saline water), gaseous phase (natural gases), and viscous phase (black and dense bitumen with approximately 8° API).

In normal sandstone, sand grains are in grain-to-grain contact but oil sand is thought to have no grain contact due to the surrounding of individual grains by fines (typically silt and clay constituents) with a water envelope and/or a bitumen film. The remaining void space might be filled with water, bitumen, and gas in various proportions. The sand material in the formation is represented by quartz and clays (99% by weight), where fines content is approximately 30% by weight; the clay content and clay size are important factors which affect the bitumen content.

2.3.2 Properties

Oil sand properties that are of general interest are bulk density, porosity, and permeability. Porosity is, by definition, the ratio of the aggregate volume of the interstices between the particles to the total volume and is expressed as a percentage. High-grade oil sand usually has porosity in the range from 30% to 35% that is somewhat higher than the porosity (5–25%) of most deposit sandstone. The higher porosity of the oil sand has been attributed to the relative lack of mineral cement (chemically precipitated material that binds adjacent particles together and gives strength to the sand, which in most sandstone occupies a considerable amount of what was void space in the original sediment).

Permeability is a measure of the ability of a sediment or rock to transmit fluids. It is, to a major extent, controlled by the size and shapes of the pores as well as the channels (throats) between the pores; the smaller the channel; the more difficult it is to transmit the deposit fluid (water, bitumen). Fine-grained sediments invariably have a lower permeability than coarse-grained sediments, even if the porosity is equivalent. It is not surprising that the permeability of the bitumen-free sand from the Alberta deposits is quite high. On the other hand, the bitumen in the deposits, immobile at formation temperatures (approximately 4°C, 40°F) and pressures, actually precludes any significant movement of fluids through the sands under unaltered formation conditions.

For the Canadian oil sands, bitumen contents from 8% to 14% by weight may be considered as normal (or average). Bitumen contents above or below this range have been ascribed to factors that influence impregnation of the sand with the bitumen (or the bitumen precursor). There are also instances where bitumen contents in excess of 12% by weight have been ascribed to gravity settling during the formative stages of the bitumen. Bitumen immobility then prevents further migration of the bitumen itself or its constituents.

In terms of specific geological and geochemical aspects of the formation, the majority of the work has, again, been carried out on the Athabasca deposit. Attention has repeatedly been focused on the variation in physical properties of crude oil produced in multiple-zone fields or in some instances within a single. Of all the properties, specific gravity, or American Petroleum Institute gravity, is the variable usually observed to be changing. This may simply reflect compositional differences, such as the gasoline content or asphalt content, but analysis may also show significant differences in sulfur content or even in the proportions of the various hydrocarbon types (Speight, 2005a,b, 2007, 2009).

In the more localized context of the Athabasca deposit, inconsistencies arise presumably because of the lack of mobility of the bitumen at formation temperature (approximately 4°C, 39°F). For example, the proportion of bitumen in the oil sand increases with depth within the formation. Furthermore, the proportion of the nonvolatile asphaltene constituents or the nonvolatile asphaltic fraction (asphaltene constituents plus resins) in the bitumen also increases with depth within the formation, leading to reduced yields of distillate from the bitumen obtained from deeper parts of the formation. In keeping with the concept of higher proportions of asphaltic fraction (asphaltene constituents plus resins), variations (horizontal and vertical) in bitumen properties have been noted previously, as have variations in sulfur content, nitrogen content, and the metals content (Speight, 2005a, 2007, 2009).

The richer oil sand deposits occur toward the base of the formation. However, the bitumen is generally of poorer quality than the bitumen obtained from near the top of the deposit insofar as the proportions of nonvolatile coke-forming constituents (asphaltene constituents plus resin constituents) are higher (with increased proportions of nitrogen, sulfur, and metals) near the base of the formation. The determining factor is site specificity (Speight, 2005b).

The bitumen content of the oil sand of the United States varies from 0% to as much as 22% by weight. There are, however, noted relationships between the bitumen, water, fines, and mineral contents for the Canadian tar sands. Similar relationships may also exist for the US oil sands, but an overall lack of study has prevented the uncovering of such data.

2.4 PROPERTIES OF OIL SAND BITUMEN

Oil sand bitumen is a naturally occurring material that is frequently found filling pores and crevices of sandstone, limestone, or argillaceous sediments or deposits where the permeability is low. Bitumen is reddish brown to black in color and occurs as a semisolid or solid that can exist in nature with no mineral impurity or with mineral matter contents that exceed 50% by weight.

Oil sand bitumen is extremely susceptible to oxidation by aerial oxygen. The oxidation process can change the properties (such as viscosity) to such an extent that precautions need to be taken not only in the separation of the bitumen from the sand but also during storage (Wallace, 1988; Wallace et al., 1988a,b).

Table 2.2 Distillation Data (Cumulative % by Weight Distilled) for Bitumen and Crude Oil (Speight, 2007)

Cut Point		Cumulative % by Weight Distilled		
°C	°F	Athabasca	PR Spring	Leduc (Canada)
200	390	3	1	35
225	435	5	2	40
250	480	7	3	45
275	525	9	4	51
300	570	14	5	
325	615	26	7	
350	660	18	8	
375	705	22	10	
400	750	26	13	
425	795	29	16	
450	840	33	20	
475	885	37	23	
500	930	40	25	
525	975	43	29	
538	1,000	45	35	
538+	1,000+	55	65	

Bitumen is a high boiling resource with little, if any, of the constituents boiling below 350°C (660°F). In fact, the boiling range may be approximately equivalent to the boiling range of an atmospheric residuum that is produced as a refinery product (Tables 2.2 and 2.3).

There are wide variations in the bitumen saturation of oil sand (0% by weight to 18% by weight bitumen) even within a particular deposit, and the viscosity is particularly high. Of particular note is the variation of the density of Athabasca bitumen with temperature and the maximum density difference between bitumen and water occurring at 70–80°C (160–175°F), hence the choice of the operating temperature of the hot water bitumen extraction process.

The character of bitumen can be assessed in terms of API gravity, viscosity, and sulfur content (Table 2.4). Properties such as these help the refinery operator to gain an understanding of the nature of the material that is to be processed. Thus, initial inspection of the feedstock (conventional examination of the physical properties) is necessary. From this, it is possible to make deductions about the most logical means of refining.

Feedstock	Gravity (API)	Sulfur (wt%)	Nitrogen (wt%)	Nickel (ppm)	Vanadium (ppm)	Asphaltene (heptane) (wt%)	Carbon Residue (Conradson wt%)
Arabian Light, >650°F	17.7	3.0	0.2	10.0	26.0	1.8	7.5
Arabian Light, >1050°F	8.5	4.4	0.5	24.0	66.0	4.3	14.2
Arabian Heavy, > 650°F	11.9	4.4	0.3	27.0	103.0	8.0	14.0
Arabian Heavy, >1050°F	7.3	5.1	0.3	40.0	174.0	10.0	19.0
Alaska, North Slope, >650°F	15.2	1.6	0.4	18.0	30.0	2.0	8.5
Alaska, North Slope, >1050°F	8.2	2.2	0.6	47.0	82.0	4.0	18.0
Lloydminster (Canada), >650°F	10.3	4.1	0.3	65.0	141.0	14.0	12.1
Lloydminster (Canada), >1050°F	8.5	4.4	0.6	115.0	252.0	18.0	21.4
Kuwait, >650°F	13.9	4.4	0.3	14.0	50.0	2.4	12.2
Kuwait, >1050°F	5.5	5.5	0.4	32.0	102.0	7.1	23.1
Tia Juana, >650°F	17.3	1.8	0.3	25.0	185.0		9.3
Tia Juana, >1050°F	7.1	2.6	0.6	64.0	450.0		21.6
Taching, >650°F	27.3	0.2	0.2	5.0	1.0	4.4	3.8
Taching, >1050°F	21.5	0.3	0.4	9.0	2.0	7.6	7.9
Maya, >650°F	10.5	4.4	0.5	70.0	370.0	16.0	15.0

Table 2.3 Properties of Selected Atmospheric (>650°F) and Vacuum (>1050°F) Residua (Speight, 2007)

Table 2.4 Properties of Bitumen Isolated from Various Utah Tar-Sand Deposits (Oblad et al., 1997)				
Properties	Whiterocks	Asphalt Ridge	PR Spring	Sunnyside
S.G. (15/15°C)	0.98	0.985	1.005	1.015
API gravity	12.9	12.1	9.3	7.9
CCR, wt%	9.5	13.9	14.17	15.0
Pour point, °C	54	47	46	75
Ash, wt%	0.8	0.04	3.3	2.4
Viscosity, at 70°C (Pa s)	4,825	5,050	47,000	173,000
Molecular weight	635	426	670	593
SARA, wt%				
Saturates	35.7	39.2	33.4	20.0
Aromatics	7.0	9.0	3.6	15.1
Resins	54.5	44.1	43.8	36.8
Asphaltenes[a]	2.9	6.8	19.3	23.6
Elemental analysis, wt% (dry, ash-free basis; oxygen calculated by difference)				
C	85.0	85.2	84.7	83.3
H	11.4	11.7	11.2	10.8
N	1.3	1.0	1.3	0.7
S	0.4	0.6	0.5	0.6
O	1.6	1.1	1.8	4.4
H/C	1.56	1.65	1.60	1.56
Distillation cuts, wt%				
Volatilities (<538°C)	46.6	53.5	45.4	40.9
<204°C	0.5	1.3	0.4	0.6
204–344°C	7.4	11.8	8.2	7.8
344–538°C	38.7	40.4	36.8	32.5
>538°C	53.4	46.5	54.6	59.1
[a]Pentane insolubles.				

In the context of the Athabasca deposit, compositional inconsistencies arise because of the lack of mobility of the bitumen at formation temperature (approximately 4°C, 39°F). For example, the proportion of bitumen in the oil sand increases with depth within the formation. Furthermore, the proportion of destructive distillate from the bitumen also decreases with depth within the formation, leading to reduced yields of distillate from the bitumen obtained from deeper parts of the formation. Variations (horizontal and vertical) in bitumen properties have been noted previously, as have variations in sulfur content, nitrogen content, and metals content.

Obviously, the richer oil sand deposits occur toward the base of the formation. However, the bitumen is generally of poorer quality than the bitumen obtained from near the top of the deposit insofar as the proportions of nonvolatile coke-forming constituents (asphaltenes plus resins) are higher (with increased proportions of nitrogen, sulfur, and metals) near the base of the formation.

2.4.1 Composition

2.4.1.1 Elemental Composition

The elemental composition (ultimate composition) of oil sand bitumen has been widely reported. However, the data suffer from the disadvantage that identification of the source is very general (i.e., Athabasca bitumen) or analysis is quoted for separated bitumen that may have been obtained by, say, hot water separation or solvent extraction and may therefore not represent the total bitumen on the sand.

However, of the available data, the elemental composition of oil sand bitumen is generally constant and falls into the same narrow range as for petroleum (Speight, 2007, 2009). In addition, the ultimate composition of the Alberta bitumen does not appear to be influenced by the proportion of bitumen in the oil sand or by the particle size of the oil sand minerals.

Of the data that are available for bitumen, the proportions of the elements vary over fairly narrow limits:

Carbon, 83.4–0.5%
Hydrogen, 10.4–0.2%
Nitrogen, 0.4–0.2%
Oxygen, 1.0–0.2%
Sulfur, 5.0–0.5%
Metals (Ni and V), >1,000 ppm

The major exception to these narrow limits is the oxygen content of heavy oil and especially bitumen, which can vary from as little as 0.2% to as high as 4.5%. This is not surprising, since when oxygen is estimated by difference the analysis is subject to the accumulation of all of the errors in the other elemental data. In addition, bitumen is susceptible to aerial oxygen and the oxygen content is very dependent upon the sample history. For example, it has been shown that bitumen is susceptible to oxidation during separation from the sand as well as when the bitumen samples are not protected by a blanket of dry nitrogen gas (Moschopedis and Speight, 1974a,b, 1975, 1976). The end result of the oxidation process is an increase in viscosity.

Therefore, bitumen cannot be adequately or sufficiently defined using viscosity or any other property that is susceptible to changes during the course of the sample's history. Similarly, a sample that is identified as oil sand bitumen without any consideration of the sample history will lead to an erroneous diagnosis of the sample.

2.4.1.2 Chemical Composition

The precise chemical composition of bitumen is, despite the large volume of work performed in this area, largely speculative (Strausz and Lown, 2003; Speight, 2007, 2009). In very general terms (and as observed from elemental analyses), bitumen is a complex mixture of (1) hydrocarbons, (2) nitrogen compounds, (3) oxygen compounds, (4) sulfur compounds, and (5) metallic constituents. However, this general definition is not adequate to describe the composition of bitumen as it relates to its behavior.

In addition, the Athabasca oil sand is extremely heterogeneous with respect to physical deposit characteristics such as geometry, component distribution, porosity, and permeability; mineralogy and mineral chemistry; aqueous fluid distribution and chemistry; and the distribution and chemistry of bitumen. Variations in these properties appear to be interrelated and reflect the dynamic and complex depositional history as well as postdepositional oil alteration processes (Fustic et al., 2006).

2.4.2 Fractional Composition

Physical methods of fractionation of oil sand bitumen can also produce the four generic fractions: saturates, aromatics, resin constituents, and asphaltene constituents. However, for oil sand bitumen, fractionation shows that bitumen contains high proportions of *asphaltene constituents* and *resin constituents*, even in amounts up to 50% w/w (or higher) of the bitumen with much lower proportions of saturates and aromatics than with petroleum or heavy oil. Much of the focus has been on the asphaltene fraction because of its high sulfur content and high coke-forming propensity. The use of composition data to model feedstock behavior during refining is becoming increasingly important in refinery operations.

In addition, the presence of ash-forming metallic constituents, including such organometallic compounds as vanadium and nickel, is also a distinguishing feature of bitumen.

Bitumen composition varies depending on the locale within a deposit due to the immobility of the bitumen at formation conditions. However, whether or not this is a general phenomenon for all oil sand deposits is unknown. The available evidence is specific to the Athabasca deposit. For example, bitumen obtained from the northern locales of the Athabasca deposit (Bitumount, Mildred-Ruth Lakes) has a lower amount (approximately 16% by weight to approximately 20% by weight) of the nonvolatile asphaltene fraction than the bitumen obtained from southern deposits (Abasand, Hangingstone River; approximately 22% by weight to approximately 23% by weight asphaltenes). In addition, other data indicate that there is also a marked variation of asphaltene content in the oil sand bitumen with depth in the particular deposit.

Oxidation of bitumen with common oxidizing agents, such as acid and alkaline peroxide, acid dichromate, and alkaline permanganate, occurs. Oxidation of bitumen in solution, by air, and in either the presence or absence of a metal salt, also occurs.

Thus, changes in the fractional composition can occur due to oxidation of the bitumen during separation, handling, or storage. Similar chemical reactions (oxidation) will also occur in heavy oil, giving rise to the perception that the sample is bitumen. However, it is likely that variability of the bitumen, as retrieved from the deposit, may be a function of several parameters, including presence or absence of a water leg, the continuity of the deposit column, sedimentary facies, water chemistry (salinity), and mineralogy.

Bitumen composition can be correlated with viscosity measurements and the geology of the host rocks. The results obtained to date (Fustic et al., 2006; McPhee and Ranger, 1998; Speight, 1979, 1990) indicate that the bitumen is heterogeneous on a deposit thickness scale and while a close relationship exists between bitumen composition and viscosity, implying bitumen properties may, to some extent, be predictable, on the other hand the relationship between bitumen composition and properties may remain speculative taking into account the other factors already discussed.

2.4.3 Properties

The specific gravity of bitumen shows a fairly wide range of variation. The largest degree of variation is usually due to local conditions that affect material lying close to the faces, or exposures, occurring in surface oil sand deposits. There are also variations in the specific gravity of the bitumen found in beds that have not been exposed to weathering or other external factors.

A very important property of the Athabasca bitumen (which also accounts for the success of the hot water separation process) is the variation of bitumen density (specific gravity) of the bitumen with temperature. Over the temperature range 30–130°C (85–265°F), the bitumen is lighter than water; hence (with aeration) floating of the bitumen on the water is facilitated and the logic of the hot water process is applied.

The API gravity of known US oil sand bitumen ranges downward from about 14° API (0.973 specific gravity) to approximately 2° API (1.093 specific gravity). Although only a general relationship exists between density (gravity) and viscosity, very-low-gravity bitumen generally has very high viscosity. For instance, bitumen with a gravity of 5° or 6° API can have viscosity up to 5,000,000 cP. Elements related to API gravity are viscosity, thermal characteristics, pour point, hydrogen content, and hydrogen–carbon ratio.

It is also evident that not only are there variations in bitumen viscosity between the major Alberta deposits, but there is also considerable variation of bitumen viscosity within the Athabasca deposit and even within one location. These observations are, of course, in keeping with the relatively high proportions of asphaltenes in the denser, highly viscous samples, a trait that appears to vary not only horizontally but also vertically within a deposit.

The most significant property of bitumen is its immobility under the conditions of temperatures and pressure in the deposit. While viscosity may present an indication of the immobility of bitumen, the most pertinent representation of this property is the *pour point* (ASTM D-97; Attachment No. 3) which is the lowest temperature at which oil will pour or flow when it is chilled without disturbance under definite conditions. When used in consideration with deposit temperature, the pour point gives an indication of the liquidity of the bitumen and, therefore, the ability of the bitumen to flow under deposit conditions.

Thus, Athabasca bitumen with a pour point of 50–100°C (122–212°F) and a deposit temperature of 4–10°C (39–50°F) is a solid or near solid in the deposit and will exhibit little or no mobility under deposit conditions. Similar rationale can be applied to the Utah bitumen where pour points of 35–60°C (95–140°F) have been recorded for the bitumen with formation temperatures on the order of 10°C (50°F), also indicating a solid bitumen within the deposit and therefore immobility in the deposit. On the other hand, the California heavy oils exhibit pour points on the order of 2–10°C (35–50°F) at a deposit temperature of 35–38°C (95–100°F) and indicate that the oil is in the liquid state in the reservoir and therefore mobile.

Irrespective of the differences between the various oil sand bitumen, the factor that they all have in common is the near-solid or solid nature and therefore immobility of the bitumen in the deposit. Conversely, heavy oil in different reservoirs has the commonality of being in the liquid state and therefore having mobility in the reservoir.

In the more localized context of the Athabasca deposit, inconsistencies arise because of the lack of mobility of the bitumen at formation temperature. For example, the proportion of bitumen in the tar sand increases with depth within the formation. Furthermore, the proportion of asphaltene constituents in the bitumen or asphaltic fraction (asphaltene constituents plus resin constituents) also increases with depth within the formation, leading to reduced yields of distillate from the bitumen obtained from deeper parts of the formation. In keeping with the concept of higher proportions of asphaltic fraction (asphaltenes plus resins), variations (horizontal and vertical) in bitumen properties have been noted previously, as have variations in sulfur content, nitrogen content, and metals content.

Nondestructive distillation data (Table 2.2) show that oil sand bitumen is a high boiling material. There is usually little or no gasoline (naphtha) fraction in bitumen and the majority of the distillate falls in the gas oil lubrication distillate range (greater than 260°C; greater than 500°F). Usually, in excess of 50% by weight of oil sand bitumen is nondistillable under the conditions of the test. On the other hand, heavy oil has a considerable proportion of its constituents that are volatile below 260°C (below 500°F).

REFERENCES

Attanasi, E.D., Meyer, R.F., 2007. Natural bitumen and extra-heavy oil. In: Trinnaman, J., Clarke, A. (Eds.), 2007 Survey of Energy Resources World Energy Council, London, United Kingdom, pp. 119–143.

Blackett, R.E., 1996. Tar Sand Resources of the Uinta Basin, Utah. Open-File Report 335. Utah Geological Survey, Utah Department of Natural Resources, Salt Lake City, UT.

Carrigy, M.A., 1967. Some sedimentary features of the Athabasca oil sands. Sediment. Geol. 1, 327–352.

CDC, 1998. California Oil and Gas Fields, Volumes I, II and III. California Department of Conservation, Division of Oil, Gas, and Geothermal Resources (DOGGR), Sacramento, CA.

De Chadenedes, J.F., 1987. Surface tar-sand deposits in California: Section Exploration Histories. AAPG Special Volumes. Volume SG 25: Exploration for Heavy Crude Oil and Natural Bitumen. American Association of Petroleum Geologists, Tulsa, OK, pp. 565–570 (Section V).

Dibblee Jr., T.W., Johnston, R.L., Earley, J.W., Meyer, R.F., 1987. Geology and Hydrocarbon Deposits of the Santa Maria, Cuyama, Taft-McKittrick, and Edna Oil Districts, Coast Ranges, California: Appendix. Field Trip Guidebook. AAPG Special Volumes. Volume SG 25: Exploration for Heavy Crude Oil and Natural Bitumen. American Association of Petroleum Geologists, Tulsa, OK, pp. 685–713.

ERCB, 2011. Alberta's Energy Reserves 2010 and Supply/Demand Outlook 2011–2020. Report No. ST98-2011. Alberta Energy Resources Conservation Board, Calgary, Alberta.

Flach, P.D., 1984. Oil Sands Geology—Athabasca Deposit North. Bulletin 046. Geological Survey Department, Alberta Research Council, Edmonton, Alberta.

Fustic, M., Ahmed, K., Brough, S., Bennett, B., Bloom, L., Asgar-Deen, M., et al., 2006. Reservoir and bitumen heterogeneity in Athabasca oil sands. In: Proceedings of the 2006 CSPG-CSEG-CWLS Convention. Canadian Society of Exploration Geophysicists, Calgary, Alberta, pp. 640–652.

Gutierrez, F.J., 1981. Occurrence of Heavy Crudes and Tar Sands in Latin America. The Future of Heavy Crude and Tar Sands. McGraw-Hill, New York, NY (Chapter 12).

Hallmark, F.O., 1982. Unconventional Petroleum Resources in California. Publication No. TR 25. Division of Oil and Gas, Department of Conservation, Sacramento, CA.

Harrison, W.E., Burchfield, M.R., 1983. Tar-Sand Potential of Selected Areas in Carter and Murray Counties, South-Central Oklahoma. Special Publication 83-3. Oklahoma Geological Survey, University of Oklahoma, Norman, OK.

Keller, M., 1988. Ventura Basin Province. United States Geological Survey, US Department of the Interior, Reston, VA.

McPhee, D., Ranger, M.J., 1998. The Geological Challenge for Development of Heavy Crude and Oil Sands of Western Canada. In: Proceedings of the Seventh UNITAR International Conference on Heavy Crude and Tar Sands, Beijing, China, October 27–30.

Meyer, R.F., de Witt Jr., W., 1990. Definition and World Resources of Natural Bitumens. Bulletin No. 1944. United States Geological Survey, Reston, VA.

Meyer, R.F., Dietzman, W.D., 1981. World Geography of Heavy Crude Oils. In: Meyer, R.F., Steele, C.T. (Eds.), The Future of Heavy Crude Oil and Tar Sands McGraw-Hill, New York, NY, pp. 16.

Meyer, R.F., Duford, J.M., 1988. Resources of heavy oil and natural bitumen worldwide. In: Meyer, R.F., Wiggins, E.J. (Eds.), Geology, Chemistry, vol. 2, pp. 277–311. Also Proceedings of the Fourth UNITAR/UNDP International Conference on Heavy Crude Tar Sands. UNITAR/UNDP Information Center for Heavy Crude Tar Sands, Alberta Oil Sands and Technology Research. Authority, Edmonton, Alberta.

Meyer, R.F., Schenk, C.J., 1985. An estimate of world heavy crude oil and natural bitumen resources. In: Proceedings of the Third UNITAR/UNDP International Conference on Heavy Crude and Tar Sands, UNITAR/UNDP Information Center for Heavy Crude and Tar Sands, New York, NY, pp. 176–191.

Meyer, R.F., Attanasi, E.D., Freeman, P.A., 2007. Heavy Oil and Natural Bitumen Resources in Geological Basins of the World. Open File-Report 2007-1084. United States Geological Survey, Reston, VA.

Meyer, R.F., Fulton, P.A., Dietzman, W.D., 1984. A preliminary estimate of world heavy crude oil and bitumen resources. In: Meyer, R.F., Wynn, J.C., Olson, J.C. (Eds.), Second International Conference—The Future of Heavy Crude and Tar Sands McGraw-Hill, New York, NY.

Moschopedis, S.E., Speight, J.G., 1974a. Oxidation of bitumen in relation to its recovery from tar sand formations. Fuel 53, 21.

Moschopedis, S.E., Speight, J.G., 1974b. The recovery of bitumen by in situ oxidation. Prepr. Am. Chem. Soc. Div. Fuel Chem. 19 (2), 192.

Moschopedis, S.E., Speight, J.G., 1975. Oxidation of bitumen. Fuel 54, 210.

Moschopedis, S.E., Speight, J.G., 1976. Some aspects of bitumen oxidation. In: Proceedings of the 13th Annual Asphalt Research Meeting. US Department of Energy, Laramie Energy Research Center, Laramie, Wyoming, July.

Noger, M.C., 1999. Tar Sand Resources of Western Kentucky. Reprint 45, Series IX. Kentucky Geological Survey, University of Kentucky, Lexington, KY.

Oblad, A.G., Dahlstrom, D.A., Deo, M., Fletcher, J.V., Hanson, F.V., Miller, J.D., et al. 1997. The Extraction of Bitumen from Western Oil Sands: Final Report: November 26, 1997, DOE/MC/30256-99. United States Department of Energy, Office of Fossil Energy, Federal Energy Technology Center, Washington, DC.

Phizackerley, P.H., Scott, L.O., 1978. Major tar-sand deposits of the world. In: Chilingarian, G.V., Yen, T.F. (Eds.), Bitumens, Asphalts, and Tar Sands. Elsevier, Amsterdam, p. 57.

Speight, J.G., 1979. Geochemical Influences on Petroleum Constitution and Asphaltene Structure. American Chemical Society: Division of Geochemistry, Washington, DC.

Speight, J.G., 1990. Tar Sand. In: Speight, J.G. (Ed.), Fuel Science and Technology Handbook Marcel Dekker Inc., New York, NY. Part II (Chapters 12–16).

Speight, J.G., 1999. The Desulfurization of Heavy Oils and Residua, second ed. Marcel Dekker Inc., New York, NY.

Speight, J.G. 2001. Refining heavy feedstocks: what's new and where are we going? Proceedings, Symposium on Advances in Residue and Heavy Oil Conversion, Spring Meeting, American Institute of Chemical Engineers, Houston, Texas, April 22–26. Paper 44b.

Speight, J.G., 2005a. Natural bitumen (tar sands) and heavy oil Coal, Oil Shale, Natural Bitumen, Heavy Oil and Peat, from Encyclopedia of Life Support Systems (EOLSS), Developed Under the Auspices of the UNESCO. EOLSS Publishers, Oxford.

Speight, J.G., 2005b. Geology of natural bitumen and heavy oil resources Coal, Oil Shale, Natural Bitumen, Heavy Oil and Peat, from Encyclopedia of Life Support Systems (EOLSS), Developed Under the Auspices of the UNESCO. EOLSS Publishers, Oxford.

Speight, J.G., 2007. Chemistry and Technology of Petroleum, fourth ed. CRC-Taylor and Francis Group, Boca Raton, FL.

Speight, J.G., 2008. Synthetic Fuels Handbook: Properties, Processes, and Performance. McGraw-Hill, New York, NY.

Speight, J.G., 2009. Enhanced Recovery Methods for Heavy Oil and Tar Sands. Gulf Publishing Company, Houston, TX.

Strausz, O.P., Lown, E.M., 2003. The Chemistry of Alberta Oil Sands, Bitumens, and Heavy Oils. Alberta Energy Research Institute, Calgary, Alberta.

US Congress, 1976. Public Law FEA 76-4. United States Congress, Washington, DC.

US DOE, 2007. A technical, economic, and legal assessment of North American heavy oil, oil sands, and oil shale resources. In Response to Energy Policy Act of 2005 Section 369(p). Work Performed Under DE-FC-06NT15569. Prepared for United States Department of Energy, Office of Fossil Energy and National Energy Technology Laboratory. Prepared by Utah Heavy Oil Program Institute for Clean and Secure Energy, The University of Utah, Salt Lake City, UT, September.

USGS, 2009. An Estimate of Recoverable Heavy Oil Resources of the Orinoco Oil Belt, Venezuela. Fact Sheet 2009-3028. United States Geological Survey, Reston, VA, October.

Wallace, D. (Ed.), 1988. A Review of Analytical Methods for Bitumens and Heavy Oils. Alberta Oil Sands Technology and Research Authority, Edmonton, Alberta.

Wallace, D., Starr, J., Thomas, K.P., Dorrence, S.M., 1988a. Characterization of Oil sand Resources. Report on the Activities Concerning Annex 1 of the US–Canada Cooperative Agreement on Tar Sand and Heavy Oil. Alberta Oil Sands Technology and Research Authority, Edmonton, Alberta, pp. 3 and 4, Appendix C.

Wallace, D., Starr, J., Thomas, K.P., Dorrence, S.M., 1988b. Characterization of Oil Sand Resources. Report on the Activities Concerning Annex 1 of the US–Canada Cooperative Agreement on Tar Sand and Heavy Oil. Alberta Oil Sands Technology and Research Authority, Edmonton, Alberta, p. 12.

Werner, M.R. 1987. West Sak and Ugnu sands: Low-gravity oil zones of the Kuparuk River area, Alaskan North Slope. In, Alaskan North Slope Geology: Bakersfield and Anchorage. I. Tailleur and P. Weimer (Editors). The Pacific Section, Society for Sedimentary Geology (SEPM) and The Alaska Geological Society. Page 109–118.

CHAPTER 3

Oil Sand Mining

3.1 INTRODUCTION

The bitumen occurring in oil sand deposits poses a major recovery problem. The material is notoriously immobile at formation temperatures and must therefore require some stimulation (usually by thermal means) in order to ensure recovery. Alternately, proposals have been noted which advocate bitumen recovery by solvent flooding or by the use of emulsifiers. There is no doubt that with time one or more of these functions may come to fruition, but, for the present, the commercial operations which involve recovery of bitumen from shallower deposits rely on a mining technique.

Approximately 20% w/w of the economically recoverable oil sands bitumen reserves are close enough to the surface to make mining feasible. Muskeg and overburden are first removed to expose the oil sands and are stockpiled for use in reclamation.

Mining of the oil sands involves excavation of the bitumen-rich sand using open pit mining methods. This is the most efficient method of extraction when there are large deposits of bitumen with little overburden. In situ methods involve processing the oil sand deposit so that the bitumen is removed while the sand remains in place. These methods are used for oil sands that are too deep to support surface mining operations to an economical degree. Eighty percent of the resource in Northern Alberta lies deep below the surface.

Such a procedure is often referred to as *oil mining*. This is the term applied to the surface or subsurface excavation of petroleum-bearing formations for subsequent removal of the heavy oil or bitumen by washing, flotation, or retorting treatments. Oil mining also includes recovery of heavy oil by drainage from reservoir beds to mine shafts or other openings driven into the rock, or by drainage from the reservoir rock into mine openings driven outside the oil sand but connected with it by bore holes or mine wells.

Oil mining is not new. Mining of petroleum and bitumen occurred in the Sinai Peninsula, the Euphrates valley, and in Persia prior to 5,000 BC. In addition, subsurface oil mining was used in the Pechelbronn oil field in Alsace, France, as early as 1735. The latter involved the sinking of shafts to the reservoir rock, only 100–200 ft (30–60 m) below the surface and the excavation of the oil sand in short drifts driven from the shafts. These oil sands were hoisted to the surface and washed with boiling water to release the bitumen. The drifts were extended as far as natural ventilation permitted. When these limits were reached, the pillars were removed and the openings filled with waste. This type of mining continued at Pechelbronn until 1866, when it was found that oil could be recovered from deeper, and more prolific, sands by letting it drain in place through mine openings with no removal of sand to the surface for treatment (http://www.musee-du-petrole.com/page14.htm). Nevertheless, mining for petroleum is a new challenge facing the petroleum industry.

Even though estimates of the recoverable oil from the Athabasca deposits are only of the order of 27 billion barrels (27×10^9 bbl) of synthetic crude oil (representing <10% of the total in-place material), this is, for the Canadian scenario, approximately six times the estimated volume of recoverable conventional crude oil. In addition, the comparative infancy of the development of the alternative options virtually ensured the adoption of the mining option for the first two (and even later) commercial ventures.

Underground mining options have also been proposed but for the moment have not been developed because of the fear of collapse of the formation onto any operation/equipment. This particular option should not, however, be rejected out of hand because a novel aspect or the requirements of the developer (which removes the accompanying dangers) may make such an option acceptable. Currently, bitumen is recovered commercially from tar and deposits by a mining technique. This produces oil sand that is sent to the processing plant for separation of the bitumen from the sand prior to upgrading.

Engineering a successful oil mining project must address a number of factors because there must be sufficient recoverable resources, the project must be conducted safely, and the project should be engineered to maximize recovery within economic limits. The use of a reliable screening technique is necessary to locate viable candidates. Once the candidate is defined, this should be followed by an exhaustive literature search covering the local geology, drilling, production, completion, and secondary and tertiary recovery operations.

The properties of the deposit which can affect the efficiency of bitumen production by mining technology can be grouped into three classes:

1. *Primary properties*, that is, those properties that have an influence on the fluid flow and fluid storage properties and include rock and fluid properties, such as porosity, permeability, wettability, crude oil viscosity, and pour point.
2. *Secondary properties*, that is, those properties that significantly influence the primary properties, including pore size distribution, clay type, and content.
3. *Tertiary properties*, that is, those other properties that mainly influence the oil production operations (fracture breakdown pressure, hardness, and thermal properties) and the mining operations (e.g., temperature, subsidence potential, and fault distribution).

There are also important rock mechanical parameters of the formation in which a tunnel is to be mined and from where all oil mining operations will be conducted. These properties are mostly related to the mining aspects of the operations and not all are of equal importance in their influence on the mining technology. Their relative importance also depends on the individual reservoir.

Many of the candidate reservoirs for application of improved oil mining are those with high oil saturation resulting from the adverse effects of reservoir heterogeneity. Faulting, fracturing, and barriers to fluid flow are features that cause production of shallow reservoirs by conventional methods to be inefficient. Production of heterogeneous reservoirs by underground oil production methods requires consideration of the manner in which fractures alter the flow of fluids.

In addition, mining methods should be applied in reservoirs that have significant residual oil saturation and have reservoir or fluid properties that make production by conventional methods inefficient or impossible. The high well density in improved oil mining usually compensates for the inefficient production caused by reservoir heterogeneity. However, close well spacing can also magnify the deleterious effects of reservoir heterogeneity. If a high-permeability streak exists with a lateral extent that is less than the interwell spacing of conventional wells but is comparable to that of improved oil mining, the channeling is more unfavorable for the improved oil mining method.

However, overburden depth alone is not a true indicator of whether an area is capable of sustaining an economically viable surface mining operation. The ore thickness, grade, clay content, and the extent of reject zones are also important parameters to be considered in the economic evaluation of a potential oil sands mining project. The thickness of overburden, ore, and center reject can be combined to give an overburden-ore ratio (thickness of overburden plus reject zones, divided by the ore thickness), which can be used as an economic indicator of the cost of delivering a unit of ore to the extraction plant. The bitumen content (grade) and clay content give an indication of the amount of bitumen that can be recovered from the unit of ore. Therefore, this is an indication of the value of that unit. It is not unusual to use only the bitumen content to define the expected processability of oil sands.

3.2 OIL SAND MINING

Since the 1920s, open pit mining has been central to oil sands development. Although less than 10% of the Athabasca Oil Sands deposit can be mined using the surface mining technique. Surface mining is the mining method which is currently being used by Suncor Energy (formerly GCOS) and Syncrude Canada Limited to recover oil sand from the ground.

GCOS was originally owned by Sun Oil Company of Marcus Hook, Pennsylvania. The Syncrude Project is a Joint Venture undertaking among Canadian Oil Sands Partnership #1, Imperial Oil Resources,

Mocal Energy Limited, Murphy Oil Company Ltd, Nexen Oil Sands Partnership, Sinopec Oil Sands Partnership, and Suncor Energy Oil and Gas Partnership, as the project owners, and Syncrude as the project operator.

Originally, oil sands were mined using draglines to excavate the face of the formation. Bucket-wheel excavators and long conveyor belts moved the raw bitumen to on-site processing facilities. This method has been replaced by shovel-and-truck mining, which gives greater flexibility. Surface mining can be used in mineable oil sand areas which lie under 250 ft or less of overburden material. The first large-scale commercial operation used by Suncor Energy introduced bucket-wheel excavators from the coal mining industry when Suncor opened in 1967. Syncrude Canada Limited opened in 1978 and introduced draglines with 360-ft booms.

Both Suncor and Syncrude suffered from start-up problems. It took several years for reasonably stable production operations to be established. Early supply costs are estimated to have been $35 per barrel or more (dollars of the day). Substantial reductions in costs have been achieved through continual process improvement, but more dramatically through two major innovations in the 1990s. First, there was a move toward replacing the draglines and bucket-wheel reclaimers with more flexible, robust, and energy efficient trucks and power shovels. Second, hydrotransport systems were introduced to replace the conveyor belts used to transport oil sands to the processing plant.

For hydrotransport, the oil sands ore is mixed with heated water (and chemicals in some cases) at the ore preparation plant to create oil sands slurry that is pumped via pipeline to the extraction plant. Hydrotransport preconditions the ore for extraction of crude bitumen and improves energy efficiency and environmental performance compared to conveyor systems.

In the past decade, attention has been directed toward maintaining stable production by minimizing unplanned maintenance, which can significantly reduce production capabilities and increase operating costs.

Since 1997, operating costs for Suncor and Syncrude have generally been in the range of $12–18 per barrel with variations largely due to natural gas price volatility, planned and unplanned maintenance turnaround costs, and project start-up costs related to expansions.

Other projects such as the Athabasca Oil Sands Project (AOSP), Shell Canada Limited (Shell), Western Oil Sands Inc., and Chevron Canada Limited (Chevron) joint venture, had not yet achieved steady-state operations. The potential exists to lower operating costs for integrated mining and upgrading to reach and maintain production costs below $10 per barrel.

The first step in surface mining is the removal of overburden and the muskeg, a water-soaked area of decaying plant material that is 3–10 ft thick and lies on top of the overburden. First the muskeg must be drained of its water content before it is removed. Overburden, which is used to build dams and dykes around the mine, is a layer of clay, sand, and silt which lies directly above the oil sands deposit. After all of the overburden is removed, the oil sand is exposed and can be mined.

The equipment employed at an *oil sand mine* is a combination of mining equipment and an on-site transportation system that may (currently) either be conveyor belts and/or large trucks (see also below). The shovel scoops up the oil sand and dumps it into a heavy hauler truck. The heavy (400-ton) hauler truck takes the oil sand to a conveyor belt which transports the oil sand from the mine to the extraction plant. Presently, there are extensive conveyor belt systems that transport the mined oil sand from the recovery site to the extraction plant. With the development of new technologies, the conveyors have been replaced by hydrotransport technology, which is a combination of ore transport and preliminary extraction.

The equipment must be durable and strong enough to withstand extreme climate and abrasive oil sand. Mining never stops; the trucks and other equipment work day and night, every day of the year. Planning is an essential and continuous part of the process. After the bituminous sand has been recovered using the truck and shovel method, it is mixed with water and caustic soda to form a slurry and is pumped along a pipeline to the extraction plant.

In a highly fractured formation with low matrix permeability, the fluid conductivity of the fracture system may be many times that of the matrix rock. In a highly fractured reservoir with low matrix permeability and reasonably high porosity, the fracture system provides the highest permeability to the flow of oil but the matrix rock contains the greater volume of the oil in place. The rate of the flow of oil from the matrix rock into the fracture system, the extent and continuity of the fracture system, and the degree to which the production wells effectively intersect the fracture system determine the production rate. Special consideration must be given to these factors in predicting production rates in fractured reservoirs. Under favorable circumstances, higher production rates may be achieved in fractured reservoirs by improving mining methods than in less heterogeneous reservoirs. Other reservoirs that are good candidates for oil mining are those that are shallow and which have high oil saturation, have a nearby formation that is competent enough to support the mine, and cannot be efficiently produced by conventional methods.

Surface mining is the mining method that is currently being used by Suncor Energy and Syncrude Canada Limited to recover oil sand from the ground. Surface mining can be used in mineable oil sand areas which lie under 75 m (250 ft) or less of overburden material. Only 7% of the Athabasca Oil Sands deposit can be mined using the surface mining technique, as the other 93% of the deposit has more than 75 m of overburden. The other 93% will have to be mined using different mining techniques.

The first step in surface mining is the removal of muskeg and overburden. Muskeg is a water-soaked area of decaying plant material that is 1–3 m thick and lies on top of the overburden material. Before the muskeg can be removed, it must be drained of its water content. The process can take up to 3 years to complete. Once the muskeg has been drained and removed, the overburden must also be removed. Overburden is a layer of clay, sand, and silt that lies directly above the oil sands deposit. Overburden is used to build dams and dykes around the mine and will eventually be used for land reclamation projects. When all of the overburden is removed, the oil sand is exposed and is ready to be mined.

There are two methods of mining currently in use in the Athabasca Oil Sands. Suncor Energy uses the truck and shovel method of mining whereas Syncrude uses this methodology as well as draglines and bucket-wheel reclaimers. These enormous draglines and bucket wheels are being phased out and soon will be completely replaced with large trucks and shovels. The shovel scoops up the oil sand and dumps it into a heavy hauler truck. The heavy hauler truck takes the oil sand to a conveyor belt that transports the oil sand from the mine to the extraction plant. Presently, there are extensive conveyor belt systems that transport the mined oil sand from the recovery site to the extraction plant. With the development of new technologies, these conveyors are being phased out and replaced with hydrotransport technology. Hydrotransport is a combination of ore transport and preliminary extraction. After the bituminous sands have been recovered using the truck and shovel method, it is mixed with water and caustic soda to form a slurry and is pumped along a pipeline to the extraction plant. The extraction process thus begins with the mixing of the water and agitation needed to initiate bitumen separation from sand and clay.

The advantages of hydrotransport include: (1) the breakdown of large lumps of oil sands in the ore and some separation of bitumen from oil sands as the slurry moves through the pipeline, (2) much more flexibility than with conveyor belt systems, because pipelines can follow circuitous routing and be placed on uneven terrain, and (3) low-energy extraction because of separation during hydrotransport, and as extraction plant operating temperatures can be reduced to 50°C or less. The reduced energy requirements will result in lower emissions (McColl et al., 2008).

Suncor uses hydrotransport to bring ore across the Athabasca River from its Steepbank Mine. Syncrude uses hydrotransport to bring ore to the Mildred Lake upgrader from its North Mine. At more remote mines, primary extraction occurs at the mine site. Bitumen froth is then transported to a central site by pipeline for secondary extraction and upgrading. Syncrude has remote primary extraction at its Aurora Mine, 35 km north of the Mildred Lake upgrader. Suncor has remote primary extraction at its Millennium Mine on the east side of the Athabasca River. Some of the major challenges faced by hydrotransport operations include the effects of fine solids (clays), temperature, bitumen content (oil sand grade), and average sand grain size on the preconditioning process and on pipeline friction losses.

Mine spoils need to be disposed of in a manner that assures physical stabilization. This means appropriate slope stability for the pile against not only gravity but also earthquake forces. Since return of the spoils to the mine excavations is seldom economical, the spoil pile must be designed as a permanent structure whose outline blends into the landscape. Straight, even lines in the pile must be avoided.

For oil sands reserves with less than 50 m of overburden, the Alberta Chamber of Resources does not expect that any alternative to surface mining will be developed within the following 25 years. The overburden will still need to be stripped to expose the ore body, and oil sand will still have to be mined and processed in a water-based extraction process (Alberta Chamber of Resources, 2004). In general, no revolutionary changes in oil sands mining are expected to emerge, although incremental advances are expected in several fields concerning oil sands mining. Improved materials and equipment that are more durable and better suited to the oil sands industry combined with better monitoring systems for mechanical equipment to reduce production interruptions are expected. Management systems that reduce transport and handling costs may be enhanced as for decision support and information systems to improve mine management. Some reductions in bitumen loss through primary separation and reductions in the energy intensity of the extraction process may be achieved. Finally, continued improvement in the performance of existing upgrading technologies, including increased energy efficiency, catalyst development, and reductions in hydrogen use, is likely to take place.

3.3 BITUMEN SEPARATION

In terms of bitumen separation and recovery, the hot-water process is, to date, the only successful commercial process to be applied to bitumen recovery from mined oil sand in North America (Table 3.1). Many process options have been tested with varying degrees of success, and one of these options may even supersede the hot-water process.

3.3.1 The Hot-Water Process

Oil sand, as-mined commercially in Canada, contains an average of 10–12% bitumen, 83–85% mineral matter, and 4–6% water. A film of water coats most of the mineral matter, and this property permits extraction by the *hot-water process* (Carrigy, 1963a,b; Speight and Moschopedis, 1978). Many process options have been tested with varying degrees of success, and one of these options may even supersede the hot-water process.

The process utilizes the linear and the nonlinear variation of bitumen density and water density, respectively, with temperature, so that the bitumen that is heavier than water at room temperature becomes lighter than water at 80°C (180°F). Surface-active materials in the oil

Table 3.1 Bitumen Recovery Processes (Speight, 1990, 2007, 2008, 2009)
Mining
• Surface
• Subsurface
In Situ

• Thermal
  - Steam and hot water
    • Stimulation
    • Flood
  - Combustion
    • Forward
    • Reverse: wet, dry
  - Electrical
  - Nuclear
• Nonthermal
  - Diluents
    • Miscible displacement: hydrocarbons, inert gases, carbon dioxide
    • Solvent
    • Chemical: polymer, caustic, surfactant polymer
  - Emulsification
  - Bacterial

sand also contribute to the process. The essentials of the hot-water process involve a conditioning, separation, and scavenging.

Thus, in the hot-water extraction process, the oil sand feed is introduced into a *conditioning* drum. In this step, the oil sand is heated, mixed with water, and agglomeration of the oil particles begins. The conditioning is carried out in a slowly rotating drum that contains a steam-sparging system for temperature control as well as mixing devices to assist in lump size reduction and a size ejector at the outlet end. The oil sand lumps are reduced in size by ablation and mixing action. The conditioned *pulp* has the following characteristics: (1) solids 60–85% and (2) pH 7.5–8.5.

In the conditioning step, also referred to as mixing or pulping, tar-sand feed is heated and mixed with water to form a pulp of 60% by weight to 85% by weight solids at 80–90°C (175–196°F). First the lumps of oil sand as-mined are reduced in size by ablation, that is, successive layers of lump are warmed and sloughed off revealing cooler layers. The conditioned pulp is screened through a double-layer vibrating screen. Water is then added to the screened material (to achieve more beneficial pumping conditions) and the pulp enters the separation cell through a central feed well and distributor.

The bulk of the sand settles in the cell and is removed from the bottom as tailing, but the majority of the bitumen floats to the surface and is removed as froth. A middlings stream (mostly of water with suspended fines and some bitumen) is withdrawn from approximately midway up the side of the cell wall. Part of the middlings is recycled to dilute the conditioning-drum effluent for pumping. Clays do not settle readily and generally accumulate in the middlings layer. High concentrations of clays increase the viscosity and can prevent normal operation in the separation cell. Thus, it is necessary to withdraw a drag stream to act as a purge; this is usually done at high clay concentrations but may not be as essential with a low-clay oil sand charge.

Under certain operating conditions, it may be necessary to withdraw a middlings stream to the scavenger cells (air flotation cells to recover bitumen from the drag stream). The froth from the scavenger unit(s) usually has a high mineral and water content, which can be removed by gravity settling in froth settlers after which the froth is combined with the froth from the main separation cell from the centrifuge plant for dewatering and demineralizing. Before the centrifuging operation, the froth is deaerated and naphtha added to lower the viscosity for a more efficient water and mineral removal operation.

The separation cell acts like two settlers, one on top of the other. In the lower settler the sand settles down, whereas in the upper settler the bitumen floats. The bulk of the sand in the feed is removed from the bottom of the separation cell as tailings. A large portion of the feed bitumen floats to the surface of the separation cell and is removed as froth. A middlings stream consists mostly of water with some suspended fine minerals and bitumen particles. A portion of the middlings may be returned for mixing with the conditioning-drum effluent in order to dilute the separation-cell feed for pumping. The remainder of the middlings is called the drag stream, which is withdrawn from the separation cell to be rejected after processing in the scavenger cells.

Tar-sand feed contains a certain portion of fine minerals that, if allowed to build up in concentration in the middlings, increases the viscosity and eventually disrupts settling in the separation cell. The drag stream is required as a purge in order to control the fines concentration in the middlings. The amounts of water that can enter with the feed and leave with the separation-cell tailings and froth are relatively fixed. Thus, the size of the drag stream determines the makeup water requirement for the separation cell. The separation cell is an open vessel with straight sides and a cone bottom. Mechanical rakes on the bottom move the sand toward the center for discharge. Wiper arms rotating on the surface push the froth to the outside of the separation cell where it overflows into launders for collection.

The combined froth from the separation cell and scavenging operation contains an average of about 10% by weight mineral material and up to 40% by weight water. The dewatering and demineralizing is accomplished in two stages of centrifuging; in the first stage the coarser mineral material is removed but much of the water remains. The feed then passes through a filter to remove any additional large-size mineral matter that would plug up the nozzles of the second-stage centrifuges.

The third step in the hot-water process is scavenging. Depending on the drag-stream size and composition, enough bitumen may leave the process in the drag stream to make another recovery step economical. Froth flotation with air is usually employed. The scavenger froth is combined with the separation-cell froth to be further treated and upgraded to synthetic crude oil. Tailings from the scavenger cell join the separation-cell tailings stream and go to waste. Conventional froth-flotation cells are suitable for this step.

Froth from the hot-water process may be mixed with a hydrocarbon diluent, for example, coker naphtha, and centrifuged. The Suncor process employs a two-stage centrifuging operation, and each stage consists of multiple centrifuges of conventional design installed in parallel. The bitumen product contains 1 to 2% by weight mineral (dry bitumen basis) and 5 to 15% by weight water (wet diluted basis). Syncrude also utilizes a centrifuge system with naphtha as the diluent.

The first commercial operations (GCOS/Suncor and Syncrude) used the Clark Hot-Water Extraction Process. Oil sands were mixed with hot water (70–80°C) and caustic in large rotating tumblers to begin separation of the bitumen from the sand. Slurry from the tumblers was fed into large primary separation vessels (PSVs) where the bitumen was separated from the sand by gravity. The bitumen floated to the surface of the PSVs as froth; the sand settled to the bottom. The froth was subjected to further processing (froth treatment) for water and solids removal. Froth treatment is required to minimize the amount of water and solids going to the upgrader, so at this point naphtha is added as a diluent, and the mixture enters a high-speed centrifuge to complete the cleaning/

separation. The diluted bitumen is moved to the upgrading unit while the sand and other materials that settle during the separation process are removed as "tailings slurry" for disposal in large tailings ponds.

The middlings and underflow streams from the PSVs are pumped to tailings oil recovery (TOR) vessels (a technology developed by Syncrude) to recover residual bitumen, which is returned to the PSVs. The middlings from the TOR vessels are processed in a secondary flotation plant for further bitumen recovery. Primary and secondary froth are combined, deaerated and heated, and fed to the froth treatment plant. Froth is diluted with naphtha for separation of solids in plate settlers and/or centrifuges. A naphtha recovery unit recovers naphtha from froth treatment tailings before the tailings are sent to disposal.

In a technique that is relatively new to the oil sands industry, bitumen froth is sent to a circuit that uses a process known as countercurrent decantation (McColl et al., 2008). A solvent is added which separates the remaining solids, water, and heavy asphaltene constituents in a three-stage, dual-circuit process. The process yields clean, diluted bitumen, low in contaminants, and with a viscosity that enables the bitumen to be transported by pipeline.

Recent groundbreaking work by Shell Canada has also resulted in the innovative Shell Enhance froth treatment technology that the company is using at the AOSP's Muskeg River mine. Using higher temperatures in the processing of the oil sands froth improves energy efficiency by 10%, resulting in a 40,000-ton reduction in annual greenhouse gas (GHG) emissions and decrease in water usage by 10%, compared to conventional low-temperature paraffinic froth treatment processing.

Considerable effort is underway to reduce the energy required for bitumen extraction. At its Aurora mine, opened in 2000, Syncrude installed a low-energy extraction process that operates at approximately 35°C (95°F) and is designed to consume about one-third of the energy of the traditional 80°C (176°F) process. Success in this area would result in substantial cost reductions and have considerable environmental benefits.

The US oil sands have received considerably less attention than the Canadian deposits. Nevertheless, approaches to recover the bitumen from US oil sands have been made. An attempt has been made to develop the hot-water process for the Utah sands. The process differs significantly from that used for the Canadian sands due to the oil-wet Utah sands contrasting to the water-wet Canadian sands. This necessitates disengagement by hot-water digestion in a high-shear force field under appropriate conditions of pulp density and alkalinity. The dispersed bitumen droplets can also be recovered by aeration and froth flotation.

3.3.2 The Cold-Water Process

The cold-water bitumen separation processes have been developed to the point of small-scale continuous pilot plants (Miller and Misra, 1982). The proposed *cold-water process* for bitumen separation from mined oil sand has also been recommended (Miller and Misra, 1982; Misra et al., 1981). The process uses a combination of cold water and solvent, and the first step usually involves disintegration of the oil sand charge that is mixed with water, diluent, and reagents. The diluent may be a petroleum distillate fraction such as kerosene and is added in approximately 1:1 weight ratio to the bitumen in the feed. The pH is maintained at 9–9.5 by the addition of wetting agents and approximately 0.77 kg of soda ash per ton of oil sand. The effluent is mixed with more water, and in a raked classifier the sand is settled from the bulk of the remaining mixture. The water and oil overflow the classifier and are passed to thickeners where the oil is concentrated. Clay in the tar-sand feed has a distinct effect on the process; it forms emulsions that are hard to break and are wasted with the underflow from the thickeners.

The *sand-reduction process* is a cold-water process without solvent. In the first step, the tar-sand feed is mixed with water at approximately 20°C (68°F) in a screw conveyor with a ratio of 0.75–3 t/t oil sand (the lower range is preferred). The mixed pulp from the screw conveyor is discharged into a rotary-drum screen, which is submerged in a water-filled settling vessel. The bitumen forms agglomerates that are retained by an 840-μm (20-mesh) screen. These agglomerates settle and are withdrawn as oil product. The sand readily passes through the 840-μm (20-mesh) screen and is withdrawn as waste stream. The process is called sand reduction because its objective is the removal of sand from the oil sand to provide a feed suitable for a fluid coking process; approximately 80% of the sand is removed. Nominal composition of the oil product is 58% w/w bitumen, 27% w/w mineral, and 15% w/w water.

A process called spherical agglomeration closely resembles the sand-reduction process (48–49). Water is added to oil sands and the mixture is ball-milled. The bitumen forms dense agglomerates of 75–87% w/w bitumen, 12–25% w/w sand, and 1–5% w/w water.

3.3.3 Solvent Extraction

An anhydrous solvent extraction process for bitumen recovery has been attempted and usually involves the use of a low-boiling hydrocarbon.

The process generally involves up to four steps. In the mixer step, fresh oil sand is mixed with recycle solvent that contains some bitumen and small amounts of water and mineral. Solvent-to-bitumen weight ratio is adjusted to approximately 0.5. The drain step consists of a three-stage countercurrent wash. Settling and draining time is approximately 30 min for each stage. After each extraction step, a bed of sand is formed and the extract is drained through the bed until the interstitial pore volume of the bed is emptied. From time to time, the bed is plugged with fine mineral or emulsion. In these cases, the drainage rate is essentially zero and the particular extraction stage is ineffective. The last two steps of the process are devoted to solvent recovery. Stripping of the solvent from the bitumen is straightforward. The solvent recovery from the solids holds the key to the economic success of an anhydrous process.

Although solvent extraction processes have been attempted and demonstrated for the Athabasca, Utah, and Kentucky oil sands, solvent losses influence the economics of such processes and they have not yet been reduced to commercial practice.

3.3.4 Tailings Management

One of the major problems that arise from the hot-water process is the disposal and control of the tailings. The fact is that each ton of oil sand in place has a volume of about 16 ft^3, which will generate about 22 ft^3 of tailings giving a volume gain on the order of 40%. If the mine produces about 200,000 tons of oil sand per day, the volume expansion represents a considerable solids disposal problem. Tailings from the process consist of about 49–50% by weight of sand, 1% by weight of bitumen, and about 50% by weight of water. The average particle size of the sand is about 200 microns and it is a suitable material for dike building. Accordingly, Suncor used this material to build the sand dike, but for fine sand, the sand must be well compacted.

Environmental regulations in Canada or the United States will not allow the discharge of tailings streams (1) into the river, (2) on to the surface, or (3) on to any area where groundwater domains may be contaminated. The tailings stream is essentially high in clays and contains some bitumen; hence, the current need is for tailings ponds, where some

settling of the clay occurs. In addition, an approach to acceptable reclamation of the tailings ponds will have to be accommodated at the time
of site abandonment.

The structure of the dike may be stabilized on the upstream side by
beaching. This gives a shallow slope but consumes sand during the season when it is impossible to build the dike. In remote areas such as the
Fort McMurray (Alberta) site, the dike can only be built in above-freez-
ing weather because (1) frozen water in the pores of the dike will create
an unstable layer and (2) the vapor emanating from the water creates a
fog, which can create a work hazard. The slope of the tailings dike is
about 2.5:1 depending on the amount of fines in the material. It may
be possible to build 2:1 slopes with coarser material, but steeper slopes
must be stabilized quickly by bleaching. After discharge from the hot-
water separation system, it is preferable that attempts be made to separate the sand, sludge, and water; hence, the tailings pond. The sand is
used to build dikes and the runoff that contains the silt, clay, and water
collects in the pond. Silt and some clay settle out to form sludge and
some of the water is recycled to the plant.

Another strategy being considered for fine tailings management
is Paste Technology, which rapidly dewaters the fine tailings stream to
produce a paste-like material which is still pumpable. This relatively
new technique requires synthetic flocculants to achieve rapid settling of
dense fine solid aggregates, a deep bed thickener to promote consolida-
tion in the settled solids, and dewatering channels to relieve the excess
pore pressures to form the paste. Upon discharge, the paste deposit
forms a slope and gains strength. The thickener overflow is recycled
to the plant. Research is underway to determine the parameters for its
application in the oil sands. The paste would be incorporated within the
coarse tailings disposal sites.

Other key tailings research and development initiatives proposed
for the next few years include: optimization of the composite tailings
process, reclamation of tailings deposits, management of recycle water
chemistry, and development of thickened tailings for oil sands application (McColl et al., 2008).

In summary, the hot-water separation process involves extremely
complicated surface chemistry with interfaces among various combinations of solids (including both silica sand and aluminosilicate clays),
water, bitumen, and air. The control of pH seems to be critical with the

preferred range being 8.0–8.5, which is achievable by use of any of the monovalent bases. Polyvalent cations must be excluded because they tend to flocculate the clays and thus raise the viscosity of the middlings in the separation cell.

3.4 OTHER PROCESSES

It is conceivable that the problems related to bitumen mining and bitumen recovery for the oil sand may be alleviated somewhat by the development of process options that require considerably less water in the sand/bitumen separation step. Such an option would allow a more gradual removal of the tailings ponds.

The *sand-reduction process* is a cold-water process without solvent. In the first step, the oil sand feedstock is mixed with water at approximately 20°C (68°F) in a screw conveyor in a ratio of 0.75–3 t/t oil sand (the lower range is preferred). The mixed pulp from the screw conveyor is discharged into a rotary-drum screen, which is submerged in a water-filled settling vessel. The bitumen forms agglomerates that are retained by an 840-μm (20-mesh) screen. These agglomerates settle and are withdrawn as oil product. The sand readily passes through the 840-μm (20-mesh) screen and is withdrawn as waste stream. The process is called sand reduction because its objective is the removal of sand from the oil sand to provide a feed suitable for a fluid coking process; approximately 80% of the sand is removed. The typical composition of the product is 58% by weight of bitumen, 27% by weight of mineral matter, and 15% by weight of water.

The *spherical agglomeration process* resembles the sand-reduction process. Water is added to oil sands and the mixture is ball-milled. The bitumen forms dense agglomerates of 75 to 87% by weight of bitumen, 12 to 25% by weight of sand, and 1 to 5% by weight of water.

An *oleophilic sieve process* (Kruyer, 1982, 1983) offers the potential for reducing tailings pond size because of a reduction in the water requirements. The process is based on the concept that when a mixture of an oil phase and an aqueous phase is passed through a sieve made from oleophilic materials, the aqueous phase and any hydrophilic solids pass through the sieve but the oil adheres to the sieve surface on contact. The sieve is in the form of a moving conveyor, and the oil is captured in a recovery zone; recovery efficiency is high.

An anhydrous *solvent extraction process* for bitumen recovery has been attempted and usually involves the use of a low-boiling hydrocarbon. The process generally involves up to four steps. In the mixer step, fresh oil sand is mixed with recycle solvent that contains some bitumen and small amounts of water and mineral. Solvent-to-bitumen weight ratio is adjusted to approximately 0.5. The drain step consists of a three-stage countercurrent wash. Settling and draining time is approximately 30 min for each stage. After each extraction step, a bed of sand is formed and the extract is drained through the bed until the interstitial pore volume of the bed is emptied. From time to time, the bed is plugged with fine mineral or emulsion. In these cases, the drainage rate is essentially zero and the particular extraction stage is ineffective. The last two steps of the process are devoted to solvent recovery. Stripping of the solvent from the bitumen is straightforward. The solvent recovery from the solids holds the key to the economic success of an anhydrous process.

Another aboveground method of separating bitumen from mined oil sand involves *direct heating of the oil sand* without previous separation of the bitumen (Gishler, 1949). Thus, the bitumen is not recovered as such but is an upgraded overhead product. In the process, the sand is crushed and introduced into a vessel, where it is contacted with either hot (spent) sand or with hot product gases that furnish part of the heat required for cracking and volatilization. The volatile products are passed out of the vessel and are separated into gases and (condensed) liquids. The coke that is formed as a result of the thermal decomposition of the bitumen remains on the sand, which is then transferred to a vessel for coke removal by burning in air. The hot flue gases can be used either to heat incoming oil sand or as refinery fuel. As expected, processes of this type yield an upgraded product but require various arrangements of pneumatic and mechanical equipment for solids movement around the refinery.

The Taciuk process (Alberta Taciuk process, AOSTRA Taciuk process) is an aboveground oil-shale retorting technology classified as a hot recycled solids technology. The distinguishing feature of the process is that the drying and pyrolysis of the oil sand, as well as the combustion, recycling, and cooling of spent materials and residues, occur within a single rotating multichamber horizontal retort (Figure 3.1) (Taciuk et al., 1994).

In an *improved mining process*, directional (horizontal or slant) wells are drilled into the reservoir from a mine in an underlying formation to

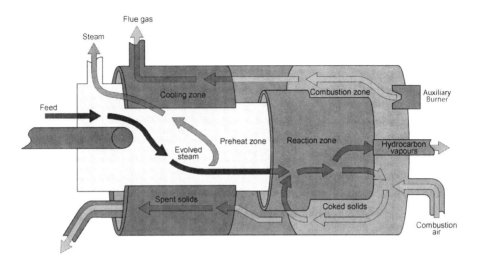

Fig. 3.1 The Taciuk retort.

drain oil by pressured depletion and gravity drainage. In the process of gravity drainage extraction of liquid crude oil, the wells are completed so that only the forces acting within the reservoir are used. The forces acting on the reservoir are left intact, perhaps maintained or increased. A large number of closely spaced wells can be drilled into a reservoir from an underlying tunnel more economically than can the same number of wells from the surface. In addition, only one pumping system is required in underground drainage, whereas at the surface each well must have a pumping system. The objective of using a large number of wells is to produce each well slowly so that the gas/oil and water/oil interfaces move toward each other efficiently. By maintaining the reservoir pressures because of forces acting on the reservoir, it is then assured that the oil production is provided by the internal forces due to gravity (the buoyancy effect) and capillary effects.

The recovery efficiency is improved by applying an enhanced oil recovery method. The production drain holes should be surveyed while drilling, and the drilling should be accomplished using mud-operated drills with a bent sub for control of direction. The drain holes must be controlled so that a network of uniformly spaced holes conforming to the data output from the computer modeling can be drilled in the production zone. The drill string should be equipped with check valves to minimize the backflow of mud during the installation of additional drill pipe. The return mud line should be equipped with a blow line to safely vent to the surface any formation gas encountered. All drilling should be accomplished working through a blowout preventer or diverter. Drill cuttings should be contained in a closed system and not allowed to encumber the mine atmosphere.

Large vertical shafts sunk from the surface are generally the means through which underground openings can be excavated. These shafts are one means of access to offer an outlet for removal of excavated rock, provide sufficient opening for equipment, provide ventilation, and allow the removal of oil and gas products during later production. These requirements plus geological conditions and oil reservoir dimensions determine the shaft size. It is expected that an access shaft will range from 8 to 20 ft in diameter.

Completing wells from a level of drifts beneath the reservoir is the most economical application of gravity drainage. In this case, only one pumping system is required rather than installing a pump in each well as is necessary in wells drilled from the surface. When wells are drilled from beneath the producing formation in an oil–water system, the developer has two options for completion. The casing may be set totally through the formation and the region opposite the oil saturation perforated to permit production. If desired, the casing may pass only through the water-saturated zone, and then a jet slotting process can be used to wash out or drill slots in the oil-saturated zone to increase the productivity of the individual well. Other shaft configurations include those for mining in weak rock and shafts for pumping drainage.

Excavation of shafts and tunnels during the mining process results in waste rock that must be stored or disposed of at the ground surface. The waste material cannot be dumped in a heap near the mine shaft as an eyesore for future generations and a source of air and water pollution for untold years. Air, water, and esthetics must be conserved not only for the present but also against reasonable future contingencies, either natural or man made.

The other aboveground method of separating bitumen from oil sand after the mining operation involves direct heating of the oil sand without previous separation of the bitumen. Thus, the bitumen is not recovered as such but is an upgraded product. Although several processes have been proposed to accomplish this, the common theme is to heat the oil sand to thermally decompose the bitumen in order to produce a volatile product with the coke remaining on the sand.

REFERENCES

Alberta Chamber of Resources, 2004. Oil Sands Technology Roadmap: Unlocking the Potential. Alberta Chamber of Resources, Edmonton, Alberta, Canada.

Carrigy, M.A. 1963a. Bulletin No. 14. Alberta Research Council, Edmonton, Alberta, Canada.

Carrigy, M.A., 1963b. The Oil Sands of Alberta. Information Series No. 45. Alberta Research Council, Edmonton, Alberta, Canada.

Gishler, P.E., 1949. Distillation of Oil from bituminous Sand. Can. J. Res. 27, 104–111.

Kruyer, J. 1982. Oleophilic Separation of Tar Sands, Oil-Water Mixtures, and Minerals. In: Proceedings of the Second International Conference on Heavy Crude and Tar Sands, February 7–17, Caracas, Venezuela.

Kruyer, J. 1983. Preprint No. 3d. Summer National Meeting, American Institute of Chemical Engineers, August 28–31, Denver, Colorado.

McColl, D., Mei, M., Millington, D., Kumar, C., 2008. Green Bitumen: The Role of Nuclear, Gasification, and CCS in Alberta's Oil Sands: Part I—Introduction and Overview. Study No. 119. Canadian Energy Research Institute, Calgary, Alberta, Canada.

Miller, J.C., Misra, M., 1982. Hot Water Process Development for Utah Tar Sands. Fuel Process. Technol. 6, 27.

Misra, M., Aguilar, R., Miller, J.D., 1981. Surface Chemistry Features in the Hot Water Processing of Utah Tar Sand. Separation Sci. Technol. 16 (10), 1523.

Speight, J.G., Moschopedis, S.E., 1978. Factors affecting bitumen recovery by the hot water process. Fuel Process. Technol. 1, 261–268.

Speight, J.G., 1990. Part II: Tar sands. In: Speight, J.G. (Ed.), Fuel Science and Technology Handbook Marcel Dekker Inc., New York, NY (Chapters 12–16).

Speight, J.G., 2007. The Chemistry and Technology of Petroleum, fourth ed. CRC, Taylor & Francis Group, Boca Raton, FL.

Speight, J.G., 2008. Synthetic Fuels Handbook: Properties, Processes, and Performance. McGraw-Hill, New York, NY (Chapter 4).

Speight, J.G., 2009. Enhanced Recovery Methods for Heavy Oil and Tar Sands. Gulf Publishing Company, Houston, TX.

Taciuk, W., Caple, R., Goodwin, S., Taciuk, G. 1994. Dry Thermal Processor. United States Patent 5,366,596. November 22.

CHAPTER 4

Nonthermal Methods of Recovery

4.1 INTRODUCTION

The vast bitumen reserves available in various parts of the world are becoming increasingly important as a secure future energy source (Speight, 2007, 2008, 2009). In fact, the heavy-oil resource is trillions of barrels, but the cumulative recovery totals to date are on the order of billions of barrels. Whether the potential and promise of oil sand bitumen is realized depends on the evolution of recovery technologies that are appropriate for the wide range of deposit and organic-phase conditions. Such technologies also need to be comparatively benign from an environmental aspect.

However, oil sand bitumen occurs in a deposit and does not flow in the deposit viscous (unless the deposit temperature is high) and, therefore, traditional methods of oil recovery, such as primary and secondary methods, are not applicable to oil sand bitumen. The route to use is, of course, deposit specific.

The majority of the oil sands reserves in Alberta are too deep for *open pit* mining. In a conventional oil reservoir, the hydrocarbon flows toward a series of producing wells and is pumped to the surface. Various secondary and tertiary recovery techniques may be employed to enhance the natural flow of fluids. By increasing the pressure in the deposit, these techniques may increase the porosity of the strata, reduce the viscosity of the fluids, or induce a driving force on the fluids. *In situ* extraction applies variations of conventional tertiary recovery techniques to the oil sands. The most common concepts employ steam, or fire, flooding technology with various stimulation techniques (Speight, 2009).

In terms of bitumen recovery from oil sand deposits, it is more traditional to apply thermal oil recovery techniques of which steam injection is the most popular. Thus, recovery efforts include thermal methods (steam floods, cyclic steam stimulation (CSS), and SAGD) as well as nonthermal methods (cold flow with sand production, cyclic solvent process (CSP), vapor-assisted petroleum extraction (VAPEX)) (Butler and Mokrys, 1991, 1995a,b). Significant improvements to the effectiveness of these methods can be achieved by developing a basic understanding of the complex displacement mechanisms and by developing new techniques for *in situ* characterization of fluid and deposit characteristics.

Development of optimal strategies for recovering these reserves requires the development of state-of-the-art deposit flow simulator(s) that incorporate techniques for coupling geomechanics and fluid flow besides accurately representing thermal, phase equilibria, and mass transfer effects. Improved recovery efficiency can also be achieved by various combinations of thermal and nonthermal processes. In addition, recent advances in drilling and production from unconsolidated sands allow development of bitumen recovery strategies that were not possible three decades ago.

Thus, the quest to produce oil sand bitumen is a global issue. However, in many cases, the resource does not resemble the conditions that are thought to be typical for heavy oil: viscous oil held in relatively permeable, shallow sands and the fields of interest have evolved to include fractured carbonates, offshore settings, and deeper, more geologically heterogeneous, heavy-oil resources. Such new settings introduce new challenges, which include highly permeable pathways and limits on deposit access, in addition to those of large oil-phase viscosity and low deposit energy.

Waterflooding for recovery of bitumen is summarily dismissed because of adverse mobility ratios. Bitumen viscosity decreases rapidly with increasing temperature (Speight, 2008, 2009); therefore, the presence of internal heat is an asset for production. In fact, the mobility of bitumen is increased at high deposit temperature, and the temperature of the surrounding environment is one of the deciding factors when considering bitumen recovery.

4.2 PRIMARY AND SECONDARY RECOVERY METHODS

There are reservoirs in the Athabasca oil sands area that permit primary recovery or cold production (Dunbar, 2009; McColl et al., 2008; McPhee and Ranger, 1998). That is, no external energy is applied to the reservoir to induce the bitumen to flow to the wellbore. This type of production technology is commonly referred to as *cold heavy-oil production with sand* (CHOPS). A significant difference between primary bitumen and conventional heavy-oil production is the amount of sand that is coproduced. Sand production from primary bitumen wells tends to be two to three times greater than that from conventional heavy-oil wells.

In addition, several operators have also been having success with application of secondary recovery techniques (water and polymer flooding) in the Wabasca area (Pelican Lake oil field) of the Athabasca Oil Sands area (Dunbar, 2009; Fossey et al., 1997). The recovered crude oil (Wabasca Heavy) is a blend of heavy-oil production that typically has 21° API gravity and 3.8% w/w sulfur (http://www.crudemonitor. ca/crude.php?acr=WH). While this crude oil comes from a reservoir with the area designated by the Alberta Government as Athabasca Oil Sands Area, it should not be assumed that the crude oil is oil sand bitumen—which is immobile in the deposit (8° API and 4.8% w/w sulfur) and cannot be recovered by conventional oil recovery methods.

The heavy, viscous nature of the bitumen means that it will not flow under the typical conditions of temperature and pressure that exist in the deposit(s). Consequently, advanced *in situ* recovery technologies have been developed that apply thermal energy to heat the bitumen and reduce its viscosity, thereby allowing it to flow to the wellbore.

The most common thermal techniques involve steam injection into the deposit using either CSS or steam-assisted gravity drainage (SAGD). Steam is injected into the oil sands zone using vertical, deviated, horizontal, or horizontal multilateral wells. The steam raises the temperature of the bitumen, thereby reducing the viscosity and increasing bitumen mobility in the deposit. As a result, the bitumen can be brought to the surface through wells using deposit pressure, gas lift, or downhole pumps.

The primary disadvantage of steam-based thermal recovery techniques is the large amount of energy and water that must be consumed for the generation of steam. A common industry's rule-of-thumb is that 1,000 standard cubic feet of natural gas is consumed for every barrel of bitumen produced; however, many projects are using much more.

The term *cold production* refers to the use of operating techniques and specialized pumping equipment to aggressively produce heavy-oil reservoirs without applying heat and is not usually applicable to bitumen deposits (Chugh et al., 2000). This encourages the associated production of large quantities of the unconsolidated uncemented sand which, in turn, results in significantly higher oil production. In contrast, conventional practices of primary production discourage sand production and result in minimized initial unit operating costs. This practice, however, may prevent many wells from achieving their maximum oil production rate and reserve potential.

The basis of cold production is that the oil production and recovery improve when sand production occurs naturally. Field production data indicates that fluid flow is more efficient when sand is produced from unconsolidated deposits. In the Elk Point Reservoir (western Canada) and Lindberg Reservoir (Arizona), sand production from wells occurs regularly (Loughead, 1992; McCaffrey and Bowman, 1991).

A cold production oil reservoir is a *solution gas drive reservoir* if the major reservoir energy for primary depletion is supplied by the release of gas from the oil and the expansion of the in-place fluids as reservoir pressure declines. The fraction of original oil in place that can be recovered by solution gas drive decreases with increasing oil viscosity.

On the other hand, there are several very large projects (>100,000 bbl per day) for heavy-oil (approximately 12° API) recovery in the Heavy Oil Belt (Faja) in Venezuela, where the main recovery method to date has been primary recovery. The recovery efficiencies are projected to be on the order of 8–15% recovery. The foamy nature of the oil has yielded initial rates of over 1,000 bbl per day, and this is not a common recovery level for oil of this API gravity. Production from this belt is expected to last for 35 years at a production rate of 600,000 bbl per day (Meyer and Attanasi, 2003). There are a few factors and technical advances that allow this heavy oil to be produced (Curtis et al., 2002).

In the case of the FAJA oil, the viscosity is low enough with the existing solution gas that the heavy oil can flow at reservoir temperatures. Second, horizontal wells up to 5,000 ft long allow production at economic rates while maintaining sufficiently low drawdown pressures to prevent extensive sand production. More complex well geometries are being drilled with several horizontal branches (multilateral wells). Third, the horizontal legs are placed precisely in the target sands using logging-while-drilling (LWD) and measurement-while-drilling (MWD) equipment, enabling more cost-effective placement of the wells. Fourth, in some locations, sand production from the unconsolidated formation is minimized using slotted liners and other sand-control methods. A low drawdown pressure in a long multilateral well can also reduce the need for significant sand control. Finally, progressive cavity pumps (PCPs) and electric submersible pumps are also employed.

The main issue for cold production is the low recovery factor for primary production. Fields are not being developed with future, secondary processes in mind. For example, wells, cement, and completions are not designed for high temperatures encountered in steam injection and other thermal recovery processes. Horizontal and fishbone wells should be drilled in the optimum location with regard to permeability, porosity, oil composition, and distances above water or below gas, and the length of the laterals.

Drilling, MWD, and LWD technologies are key enablers for this. In horizontal and multilateral wells, being able to monitor, understand, control, and ensure the flow from different sections of the well will improve production and reduce unwanted water and/or natural gas production.

Sand production is thought to be a function of (1) the absence of clays and cementation materials, (2) the viscosity of the oil, (3) the producing water cut and gas/oil ratio, and (4) the rate of pressure drawdown (Chugh et al., 2000).

The presence of clay stabilizes the sand grains and reduces sand movement. Higher viscosity oil increases the frictional drag between the oil and the sand grains, which promotes sand movement. High water or gas production inhibits sand production because gas/water is produced instead of an oil/sand mixture. Increasing the drawdown rate also promotes sand movement because of the increase in the velocity of the fluid into the wellbore and hence increased frictional drag on the sand grains. It has been reported that gross near-wellbore failure of the formation due to sand production results in excellent productivity.

The produced sand creates a modified wellbore geometry that could have several configurations including piping tubes (wormholes), dilated zones, sheared zones, or possibly cavities. Porosity in the dilated zones may increase leading to large increases in reservoir permeability. In addition, the flow of sand with the oil has the potential to reduce the frictional drag forces on the oil and result in increased productivity in the porous region. Furthermore, fines migration, which occurs during oil production, can block pore throats and reduce the number of flow paths available for the oil. Producing sand helps to eliminate many of these bottlenecks, and the dilation of the sand also creates larger pore throats that are more difficult to block.

4.2.1 Cold Heavy-Oil Production with Sand

CHOPS is used as a production approach in unconsolidated sand-stones. The process results in the development of high-permeability channels (*wormholes*) in the adjacent low-cohesive-strength sands, facil-itating the flow of oil foam that is caused by solution gas drive.

The immobile oil sand bitumen that is amenable to cold production methods is heavier than typical heavy oil (which has mobility in the res-ervoir) but lighter than the oil sand bitumen that is recovered through mining and thermal stimulation methods. Typically, cold bitumen recovery wells have productive lives of 4–10 years with 60–70% of total recovered bitumen being produced in the first 3 or 4 years. A significant level of ongoing drilling is required to maintain production.

The key benefits of the process are improved reservoir access, order-of-magnitude higher production rates (as compared to primary recov-ery), and lower production costs. The outstanding technical issues involve sand handling problems, field development strategies, wormhole plugging for water shutoff, low ultimate recovery, and sand disposal. Originally, cold production mechanisms were thought to apply only to vertical wells with high-capacity pumps. It is now believed that these mechanisms may also apply to horizontal wells and the easier flowing heavy oils.

Thus, instead of blocking sand ingress by screens or gravel packs, sand is encouraged to enter the wellbore by aggressive perforation and swabbing strategies. Vertical or slightly inclined wells (vertical to 45°) are operated with rotary PCPs (rather than reciprocating pumps), and old fields are converting to higher-capacity PCPs, giving production boosts to old wells. Productivity increases over conventional production, and a CHOPS process equate to as much as 12% to 25% of the original oil in place being recovered, rather than the 0–5% typical of primary production without sand in such cases. Finally, because massive sand production creates a large disturbed zone, the reservoir may be positively affected for later implementation of thermal processes.

The CHOPS process increases productivity because: (1) if the sand can move or is unconsolidated, the basic permeability to fluids is enhanced; (2) as more sand is produced, a growing zone of greater permeability is generated, similarly to a large-radius well which gives better production; (3) gas coming out of solution does not generate a continuous gas phase—rather, bubbles flow with the fluid and do not coalesce, but expand down-gradient, generating an *internal* gas drive, referred to as *foamy flow* (this also helps to locally destabilize the sand, sustaining the process); (4) continuous sand production means that asphaltene or fines plugging of the near-wellbore environment potentially do not occur, so there is no possibility of an effect to impair productivity; and (5) as sand is removed, the overburden weight acts to shear and destabilize the sand, helping to drive sand and oil toward the wellbore.

Typically, a well placed on CHOPS production will initially produce a high percentage of sand, greater than 20% by volume of liquids. However, this generally drops after some weeks or months. The huge volumes of sand are disposed of by slurry fracture injection or salt cavern placement or by sand placement in a landfill in an environmentally acceptable manner.

Thus, the main conditions for successful CHOPS are: (1) continuous sand failure, that is, unconsolidated sands, (2) active foamy oil, that is, sufficient gas in solution, and (3) no free water zones in the reservoir and the use of PCPs. In many cases, immobile highly viscous oil sand bitumen cannot always fulfill these criteria and application of the technology is site specific.

CHOPS is used for thin subsurface oil sand deposits (typically 1–7 m thick) in Canada, provided the oil sand is unconsolidated and the bitumen contains sufficient solution gas to power the production process. To have any natural gas in solution, the oil sand must be at least a few hundred meters deep. For example, there are a large number of CHOPS wells located near Lloydminster, Alberta. In fact, at the time of writing, CHOPS is the only commercial method for exploiting thin oil sands. CHOPS wells (by definition) require sand production. Foamy oil production may occur without sand production in other areas, such as in the FAJA belt, Venezuela. Alternatively, oil may be produced with sand but without solution gas in still other areas.

In the Cold Lake region of Alberta, CHOPS is used with vertical production wells and PCPs. In addition, the oil sand deposits in the Wabasca area of the Athabasca region and the Seal area of the Peace River region use horizontal wells to achieve comparable production rates to the CHOPS process in Cold Lake but without the production of sand on the same scale. Generally, lower viscosity is associated with lower rates of sand production. The viscosity of the bitumen in the Wabasca and Seal areas is lower than in the Cold Lake region; therefore, less sand is produced and handling costs are lower.

The surface footprint for CHOPS wells is small, only requiring space for the wellhead, a storage tank, and a small doghouse. Any produced gas is used on site to power equipment or to heat the storage tank. Because a large volume of sand is produced, pipelines cannot be used for transportation. Instead, trucks are required to move oil, water, and sand for processing or disposal. During spring break-up, the CHOPS wells in Alberta must be shut in since trucks cannot navigate the roads.

The technical challenges for CHOPS wells include a better understanding of their behavior and more predictive performance models. PCPs have increased production rates, but increased reliability and longer maintenance-free periods would improve economics. A method for water shutoff would bring some unproductive wells back to life.

4.2.2 Pressure Pulse Technology

Pressure pulse technology (PPT) is a technology which can be used to enhance the recovery rate of nonaqueous phase liquid and to reduce solids clogging in wells, permeable reactive barriers, and fractured media (CRA, 2003).

Pressure pulse flow enhancement technology is based on the discovery that large amplitude pressure pulses that are dominated by low-frequency wave energy generate enhanced flow rates in porous media. For example, in preliminary experiments in Alberta, PPT has reduced the rate of depletion, increased the oil recovery ratio, and prolonged the life of wells.

The technology uses steady, nonseismic pulse vibrations (e.g., 15 pulses per minute) that generate a low-velocity wave effect to encourage flow of oils and small solid particles. It is effective in geologic formations exhibiting elastic properties, such as unconsolidated sediments and sedimentary rocks. It must be applied in a downhole manner in order to be effective. It has been used by the oil industry to improve oil recovery from otherwise exhausted reserves for many years. Also, it has been found that very large amplitude pressure pulses applied for 5–30 h to a blocked producing well can reestablish economic production in a CHOPS well for many months and even years. Pulsing has been applied in injector wells for improving the efficiency of waterflood patterns and has shown indications of increased oil production and decreased water cut. Additional potential applications include improving the effectiveness of matrix acidizing and diversion (Halliburton, 2004; Mok, 2004).

The mechanism by which PPT works is to generate a porosity dilation wave (a fluid displacement wave similar to a tidal wave); this generates pore-scale dilation and contraction so that oil and water flow into and out of pores, leading to periodic fluid accelerations in the pore throats. As the porosity dilation wave moves through the porous medium at a velocity of about 50–100 ft/s (40–80 m/s), the small expansion and contraction of the pores with the passage of each packet of wave energy helps unblock pore throats, increase the velocity of liquid flow, overcome part of the effects of capillary blockage, and reduce some of the negative effects of instability due to viscous fingering, coning, and permeability streak channeling.

Although a very new concept dating only since 1999 in small-scale field experiments, PPT promises to be a major adjunct to a number of oil production processes, particularly all pressure-driven processes, where it will accelerate flow rates as well as increase oil recovery factors. It is also now used in environmental applications to help purge shallow aquifers of nonmiscible phases such as oil, with about six successful case histories to date.

4.2.3 VAPEX Processes

Solvent-based methods have been developed to move oil sand bitumen. A diluent such as naphtha or light oil may be injected near the pump to reduce the viscosity of the bitumen and allow it to be more easily pumped. Alternatively, diluent may be added at the surface to facilitate pipeline transport.

The *vapor extraction process* (VAPEX process) is a nonthermal, solvent-based, relatively cold (40°C), low-pressure process in which two parallel horizontal wells are drilled with about a 15-ft vertical separation (Yazdani and Maini, 2008).

With VAPEX, a solvent such as ethane, propane, or butane, instead of steam, is injected into the deposit along with a displacement gas to mobilize the hydrocarbons in the deposit and move them toward the production well. This gives the cost advantage of not having to install steam generation facilities or purchase natural gas to produce steam. The VAPEX methodology does not require water processing or recycling, offers lower carbon dioxide emissions, and can be operated at deposit temperature with almost no loss of heat. The capital costs are estimated to be 75% of SAGD costs, while the operating costs are estimated to be 50% compared to SAGD. An additional advantage is the possibility of a reduced need of diluent, as some of the solvent diffuses into the bitumen to mobilize it. On the negative side, more wells are needed to achieve similar production rates and rates of recovery.

The physics of the process is essentially the same as for SAGD and the configuration of wells is generally similar. The process involves the injection of vaporized solvents such as ethane or propane to create a vapor chamber through which the oil flows due to gravity drainage (Butler and Jiang, 2000; Butler and Mokrys, 1991, 1995a,b). The process can be applied in paired horizontal wells, single horizontal wells, or a combination of vertical and horizontal wells. The key benefits are significantly lower energy costs, and potential for *in situ* upgrading and application to thin deposits, with bottom water or reactive mineralogy.

The process involves the injection of vaporized solvents such as ethane, propane, butane, and naphtha to create a vapor-chamber (Butler and Jiang, 2000; Butler and Mokrys, 1991, 1995a,b). The vapor travels to the oil face where it condenses into a liquid and the solvent mixed with the oil flows to the lower well (gravity drainage) and is pumped to the surface.

The VAPEX process is similar in operation to SAGD, except that a solvent such as ethane, propane, or butane, instead of steam, is injected into the deposit along with a displacement gas to mobilize the heavy oil (or bitumen) in the reservoir (or deposit) and cause movement to the production well. This offers the cost advantage of not having to install steam generation facilities or purchase natural gas to produce steam. The process method requires no water processing or recycling, offers lower carbon dioxide (CO_2) emissions, and can be operated at reservoir temperature with almost no heat loss. The capital costs are estimated to be 75% of SAGD costs, while the operating costs are estimated to be 50% compared to the SAGD process (Chapter 5). Another advantage is the possibility of reduced diluent requirement, as some of the solvent diffuses into the bitumen to mobilize it. On the other hand, more wells are needed to achieve similar production rates and recovery factors.

VAPEX offers an alternative process to recover bitumen from deposits that are not amenable to thermal processes such as deposits with bottom water and/or high water saturation, vertical fractures, low porosity, and low-thermal conductivity.

Because of the slow diffusion of gases and liquids into viscous oils, this approach, used alone, perhaps will be suited only for less viscous oils although preliminary tests indicate that there are micromechanisms that act so that the VAPEX dilution process is not diffusion rate limited and the process may be suitable for the highly viscous tar-sand bitumen (Yang and Gu, 2005).

Nevertheless, VAPEX can undoubtedly be used in conjunction with SAGD methods. As with SAGD and inert gas injection (IGI) processes, a key factor is the generation of a three-phase system with a continuous gas phase so that as much of the oil as possible can be contacted by the gaseous phases, generating the thin oil film drainage mechanism. As with IGI, vertical permeability barriers are a problem, and must be overcome through hydraulic fracturing to create vertical permeable channels, or undercut by a the lateral growth of the chamber beyond the lateral extent of the limited barrier, or "baffle."

However, as with all solvent-based processes, there is the potential for solvent losses in the deposit. These can arise, for example, due to unknown fissures in the deposit rock as well as clay lenses to which the solvent will adhere.

The best solvent for the process is uncertain, but would probably be a low-molecular-weight hydrocarbon such as ethane, propane, and butane or mixtures of these hydrocarbons or maybe with some carbon dioxide. Deposit pressure, temperature, and the tendency for unanticipated and uncontrolled deposition of asphaltene constituents (*in situ* solvent deasphalting) are the major factors in choosing the solvent. Asphaltene constituents often precipitate from the oil in the deposit, increasing the API gravity of the remaining oil, but can clog portions of the deposit. This *in situ* upgrading process may increase the value of the produced oil (Das and Butler, 1994; Luo and Gu, 2005).

The process can be applied in paired horizontal wells, single horizontal wells, or a combination of vertical and horizontal wells. The physics of the VAPEX process is essentially the same as for the SAGD process (Chapter 5) and the configuration of wells is generally similar. The key benefits are claimed to be (1) significantly lower energy costs, (2) the potential for *in situ* upgrading, and (3) application to thin deposits, with bottom water or reactive mineralogy.

In addition to issues arising from the cost of the solvent, there are the usual issues related to the interaction of the solvent with the deposit minerals—clay is known to adsorb organic solvents very strongly—and the integrity of the deposit formation and associated strata; a minor fault can cause loss of the solvent as well as environmental havoc. There is also the concern over the deposition of asphaltic material and its effect on deposit permeability. While the function of the solvents might be to extract soluble components of the bitumen, initial contact between the solvent and the oil at a low solvent-to-oil ratio will cause solubilization of the asphaltic constituents (Mitchell and Speight, 1973) with later deposition of these constituents, as the solvent-to-oil ratio increases in the later stages of the process.

Because of the slow diffusion of gases and liquids into viscous oils, this approach, used alone, may be suitable only for less viscous oils.

However preliminary tests indicate that the VAPEX dilution process may not be diffusion controlled and the process may be suitable for oil sand bitumen (Yang and Gu, 2005).

Since VAPEX is a nonthermal method, it has the potential to reduce CO_2 and other GHG emissions substantially—estimated to be as much as 85% over thermal processes. Other driving forces for VAPEX include potential for dramatically reduced water consumption compared to other extraction technologies, and related lower water-handling and surface facility costs. The technology also has significant potential economic advantages because it can be used to recover bitumen from zones that are considered too thin for traditional thermal recovery methods and offers the potential for an upgraded and higher-value product by promoting *in situ* upgrading. Paraffinic solvents cause asphaltene constituents to be precipitated and left behind in the deposit (i.e., partial upgrading occurs).

The VAPEX process, like the SAGD technology, uses horizontal well pairs to recover the bitumen. However, the process uses a hydrocarbon solvent instead of steam. This eliminates the need to burn fuel (usually natural gas) to create the steam, resulting in reduced GHG emissions. In addition, the solvent in the VAPEX process can be reused. With the VAPEX process, vaporized solvents are injected into the deposit via an upper horizontal well. The bitumen in the deposit is diluted with the solvent, which reduces its viscosity and allows the bitumen-solvent mixture to drain by gravity to the production well. On the surface, the solvents are separated from the produced bitumen and recycled.

Research carried out thus far suggests that up to 90% of the solvent used can be recovered and recycled, offering the potential for dramatic cost savings over other extraction methods. Results have also shown the quality of produced bitumen to be superior because some of the heavier fractions are left in the ground (McColl et al., 2008).

Although the VAPEX process is often considered to be analogous to the SAGD process, the major difference between the two processes is that the VAPEX process does not involve heat injection so that it operates at deposit temperature and pressure rather than at the higher temperatures of the SAGD process. It is therefore much more energy efficient than a thermal process, consuming perhaps <5% of the energy of an equivalent steam-based process, and as there are no combustion products it is cleaner than thermal processes.

The VAPEX process is not effective for all heavy-oil reservoirs and bitumen deposits because of the lack of efficient gravity drainage, particularly in thin reservoirs and deposits. A hydrocarbon gas injection process in huff-n-puff mode, that is, traditional hydrocarbon-based CSP, has been proposed and offers promise on the basis of preliminary tests (Jamaloei et al., 2012).

4.2.4 N-Solv Process

N-Solv is a patented *in situ* technology (Nenniger and Dunn, 2008) that uses warm solvent to extract bitumen from oil sands efficiently, sustainably, and economically (Ibid., 2008). Currently, N-Solv™ is preparing to demonstrate the commercial readiness of the technology at its field pilot plant located near Fort McMurray, Alberta, scheduled to start production in May 2013.

The process uses the proven horizontal well technology developed for the SAGD process but differs significantly in that it does not use any water. Instead, N-Solv uses warm propane or butane, which is injected as a vapor, and condenses underground, washing the valuable compounds out of the bitumen.

The process is expected to produce a lighter, partially upgraded, and, hence, more valuable, oil product and may recover more resource from each well at lower capital and operating costs than existing *in situ* processes. The GHG benefits associated with the technology are derived from two sources: from extraction emissions (those associated with liberating oil from the ground) and from a significant reduction in downstream upgrading requirements (and consequently, energy consumption).

The process eliminates the need for an expensive steam/water treatment plant, and asphaltene constituents and the heavy metals remain the deposit, hence 20% less of each extracted barrel becomes waste. Thus, *in situ* upgrading improves the API gravity of the produced oil (13° API vs. 8° API for SAGD). Low temperatures and pressures allow thin, shallow, and low-pressure zones to be economically attractive targets. There is no water consumption or contamination, which eliminates strain on a scarce resource, and there is no downstream contamination of water supplies.

Two improvements to the original work have made the economics of solvent extraction favorable: horizontal drilling and the development of the process. The advancement and use of horizontal drilling technology increases the contact surface area which, in turn, increases the amount and rate of oil recovered from a well pair.

In solvent extraction, the production rate is limited by the rate that the solvent diffuses into the bitumen—a subtle but key difference with the VAPEX process. The penetration rate of the solvent into the bitumen is determined by the viscosity of the bitumen.

The process involves injection of a pure, heated solvent vapor into an oil sand deposit where it condenses, delivering heat to the deposit and subsequently dissolving the bitumen, with the resulting miscible liquids flowing by gravity to a production well. The process is run at moderate temperatures and pressures and uses commercially proven horizontal well technology developed for the oil sands.

As the solvent vapor is injected into the deposit through the injection well, an extraction chamber is developed as the bitumen is dissolved and removed via the production well. The perimeter of the extraction chamber, or workface, is always cooler than the chamber itself, and this provides the mechanism for the solvent to condense at the workface. With heat delivery to the workface, high rates of bitumen dissolution into the liquid solvent are achieved. Operating temperatures are 10–50°C higher than the deposit temperature, and operating pressures are somewhat higher than the deposit pressure.

As the bitumen is dissolved into the solvent, a natural deasphalting of the bitumen occurs such that the valuable components of the bitumen are preferentially extracted while the coke-forming asphaltene constituents are safely and uniformly sequestered in the deposit. Post-extraction core analyses show that the sequestered residue contains 60–70% of the asphaltene constituents as well as much of the sulfur, heavy metals (zinc, vanadium, iron), and carbon residue that are contained in the bitumen. This allows the process to produce a partially upgraded 13°–16° API oil, while typical bitumen has an API of approximately 8°.

To achieve high rates of bitumen extraction, in addition to heating the bitumen, it is important to control noncondensable gases, such as methane, in the extraction chamber. Noncondensable gases are released from the bitumen during the extraction process and will accumulate in the chamber if no measures are taken to remove them. Noncondensable gases will blanket the extraction interface and prevent the solvent from directly condensing on the bitumen at the workface. This reduces the dissolution effectiveness of the solvent, and thus reduces the extraction rate. With the right circulation rate of solvent, the process provides the ability to minimize the concentration of noncondensable gases in the reservoir and thus sustain high production rates throughout the working life of the reservoir.

With Athabasca bitumen, a 25–30°C (45–54°F) temperature rise typically reduces the bitumen viscosity by a factor of 100. Thus, a substantial acceleration in the bitumen extraction rate is achieved with a very modest increase in temperature. To achieve the desired temperature rise, it is necessary to have a solvent purity specification, as the condensation temperature is reduced by about 5°C (9°F) for every mole percent of methane contamination. Thus, even a small amount of methane contamination in the gravity drainage chamber can greatly impair the ability of the solvent to deliver heat to the bitumen interface.

Although methane is naturally present in the *in situ* bitumen, the use of high-purity condensing solvent at moderate temperatures provides a very efficient mechanism to remove the *in situ* methane from the chamber. Thermodynamic calculations show that N-Solv will be perhaps 20 times more efficient at removing the noncondensable gas from the chamber than steam extraction processes such as SAGD.

Conversely, unheated solvent extraction processes such as the VAPEX process must add methane or another non-condensable gas to the solvent vapor in order to raise the dew point pressure to match the *in situ* pressure. However, the solvent is preferentially removed in the produced liquids, so the methane will accumulate in the chamber and eventually poison the mass transfer.

In the N-Solv process, the condensed solvent and mobilized bitumen drain down the interface and form a pool at the bottom of the chamber, where they are removed via the production well. The solvent's purity specification helps ensure that the condensed solvent has maximum practical capacity to remove noncondensable constituents from the chamber. Furthermore, N-Solv™ avoids reboiling, ensuring that the methane is efficiently removed from the chamber. This minimizes the risk of methane accumulation and consequent poisoning.

Using Athabasca bitumen at the underground test facility in *in situ* conditions has shown that the process achieves extraction rates 40–50 times faster than with the VAPEX process. The data also confirmed that selective deasphalting and noncondensable gas contamination is an effective hindrance to the process.

4.2.5 Inert Gas Injection

IGI is a technology for conventional oils in reservoirs where good vertical permeability exists, or where it can be created through propped hydraulic fracturing. It is generally viewed as a *top-down* process with nitrogen or methane injection through vertical wells at the top of the reservoirs, creating a gas–oil interface that is slowly displaced toward long horizontal production wells. As with all gravity drainage processes, it is essential to balance the injection and production volumes precisely so that the system does not become pressure driven but remains in the gravity-dominated flow regime (Mezaros et al., 1990).

High recovery ratios are achieved because, in the absence of elevated pressure gradients, a thin oil film is maintained between the gas and the water phases. The film is maintained because the sum of the oil–water and gas–oil surface tensions is always less than the water–gas surface tension. Thus, the thin film configuration is thermodynamically stable and allows the oil to drain to values far lower than the *residual oil saturation* value.

The interfaces in IGI are gravity stabilized because of the difference in phase densities, so that at slow drainage rates the interfaces remain approximately horizontal without viscous fingering. In one configuration, the horizontal wells are produced under a back-pressure equal to the pressure in the underlying water phase, so no water coning can occur. During production, if the water cut increases, the production rate is reduced so that the interface becomes stable. Alternatively, if gas is injected too quickly, gas coning can develop, and if this is observed, the gas injection rate must be reduced to sustain stability. The process is continued until the oil zone is "pinched down" to the horizontal well, achieving the high recovery ratios possible with gravity drainage methods. These principles are fundamental to all gravity-dominated processes and failure to adhere to them will drive the system into conditions of instability (such as coning and fingering).

In reservoirs with excellent vertical permeability, the bottom water zone can also be injected with water to cause the oil–water interface to rise slowly toward the production well. First implemented in Canada, IGI is used extensively in carbonate pinnacle reefs with excellent vertical permeability, and recovery ratios exceeding 80% are systematically achieved. As with all gravity drainage processes, it is generally necessary to place the horizontal wells as low in the structure as possible.

4.2.6 Chemical Enhanced Recovery

There is a renewed interest in chemical Enhanced oil recovery (EOR) (Speight, 2009) because of diminished reserves and advances in surfactant and polymer technology. Greater understanding of the chemical reactions involved has led to good results in the field (Krumrine and Falcone, 1987). Combinations of chemicals may be applied as premixed slugs or in sequence.

The choice of the method and the expected recovery depends on many considerations, economic as well as technological. Some methods are commercially successful, while others remain largely of academic interest. Only a few recovery methods have been commercially successful, such as steam injection–based processes in heavy-oil reservoirs (if the reservoir offers favorable conditions for such applications) and miscible carbon dioxide for light-oil reservoirs (Thomas, 2008). Methods for improving oil recovery, in particular those concerned with lowering the interstitial oil saturation, have received a great deal of attention both in the laboratory and in the field. From the vast amount of literature on the subject, one gets the impression that it is relatively simple to increase oil recovery beyond secondary recovery (assuming that the reservoir lends itself to primary and secondary recoveries) but this is not the case (Ibid., 2008).

4.2.7 Microbial Enhanced Oil Recovery

Microbial enhanced oil recovery (MEOR) processes involve the use of reservoir microorganisms or specially selected natural bacteria to produce specific metabolic events that lead to EOR.

The processes that facilitate oil production are complex and may involve multiple biochemical processes. Microbial biomass or biopolymers may plug high-permeability zones and lead to a redirection of the waterflood, produce surfactants which lead to increased mobilization of residual oil, increase gas pressure by the production of carbon dioxide, or reduce the oil viscosity due to digestion of large molecules.

In MEOR processes, microbial technology is exploited in oil reservoirs to improve recovery (Banat, 1995; Clark et al., 1981; Stosur, 1991). From a microbiologist's perspective, MEOR processes are somewhat akin to *in situ* bioremediation processes. Injected nutrients, together with indigenous or added microbes, promote *in situ* microbial growth and/or generation of products which mobilize additional oil and move it to producing wells through reservoir repressurization, interfacial tension/oil viscosity reduction, and selective plugging of the most permeable zones (Bryant and Lindsey, 1996; Bryant et al., 1989). Alternatively, the oil-mobilizing microbial products may be produced by fermentation and injected into the reservoir.

This technology requires consideration of the physicochemical properties of the reservoir in terms of salinity, pH, temperature, pressure, and nutrient availability (Khire and Khan, 1994a,b). Only bacteria are considered promising candidates for MEOR. Molds, yeasts, algae, and protozoa are not suitable due to their size or inability to grow under the conditions present in reservoirs. Many petroleum reservoirs have high concentrations of sodium chloride (Jenneman, 1989) and require the use of bacteria which can tolerate these conditions (Shennan and Levi, 1987). Bacteria producing biosurfactants and polymers can grow at sodium concentrations up to 8% and selectively plug sandstone to create a biowall to recover additional oil (Raiders et al., 1989).

Organisms that participate in oil recovery produce a variety of fermentation products, for example, carbon dioxide, methane, hydrogen, biosurfactants, and polysaccharides from crude oil, pure hydrocarbons, and a variety of nonhydrocarbon substrates. Organic acids produced through fermentation readily dissolve carbonates and can greatly enhance permeability in limestone reservoirs, and attempts have been made to promote anaerobic production.

The MEOR process may modify the immediate reservoir environment in a number of ways that could also damage the production hardware or the formation itself. Certain sulfate reducers can produce hydrogen sulfide, which can corrode pipeline and other components of the recovery equipment. Thus, despite numerous MEOR tests, considerable uncertainty remains regarding process performance. Ensuring success requires an ability to manipulate environmental conditions to promote growth and/or product formation by the participating microorganisms. Exerting such control over the microbial system in the subsurface is itself a serious challenge. In addition, conditions vary from reservoir to reservoir or from deposit to deposit, which calls for reservoir-specific or deposit-specific customization of the MEOR process, and this alone has the potential to undermine microbial process economic viability.

MEOR differs from chemical EOR in the method by which the enhancing products are introduced into the reservoir. Thus, in oil recovery by the *cyclic microbial method*, a solution of nutrients and microorganisms is introduced into the reservoir during injection. The injection well is then shut for an incubation period allowing the microorganisms to produce carbon dioxide gas and surfactants that assist in mobilization of the oil. The well is then opened and oil and oil products resulting from the treatment are produced. The process is repeated as often as oil can be produced from the well. Oil recovery by *microbial flooding* also involved the use of microorganisms, but in this case, the reservoir is usually conditioned by a water flush after which a solution of microorganisms and nutrients is injected into the formation. As this solution is pushed through the reservoir by water drive, gases and surfactants are formed, and the oil is mobilized and pumped through the well. However, even though microbes produce the necessary chemical reactions *in situ* whereas surface-injected chemicals may tend to follow areas of higher permeability, resulting in decreased sweep efficiency, there is need for caution and astute observation of the effects of the microorganisms on the reservoir chemistry.

The mechanism by which MEOR processes work can be quite complex and may involve multiple biochemical processes. In selective plugging approaches, microbial cell mass or biopolymers plug high-permeability zones and lead to a redirection of the waterflood. In other processes, biosurfactants are produced *in situ*, which leads to increased mobilization of residual oil. In still other processes, microbial production of carbon dioxide and organic solvents reduces the oil viscosity as the primary mechanism for EOR.

In the MEOR process, conditions for microbial metabolism are supported via injection of nutrients. In some processes, this involves injecting a fermentable carbohydrate into the reservoir. Some reservoirs also require inorganic nutrients as substrates for cellular growth or for serving as alternative electron acceptors in place of oxygen or carbohydrates.

The stimulation of oil production by *in situ* bacterial fermentation is thought to proceed by one or a combination of the following mechanisms:

1. Improvement of the relative mobility of oil to water by biosurfactants and biopolymers.
2. Partial repressurization of the reservoir by methane and carbon dioxide.
3. Reduction of oil viscosity through the dissolution of organic solvents in the oil phase.
4. Increase of reservoir permeability and widening of the fissures and channels through the etching of carbonaceous rocks in limestone reservoirs by organic acids produced by anaerobic bacteria.
5. Cleaning the wellbore region through acids and gas from *in situ* fermentation in which the gas pushes oil from dead space and dislodges debris that plugs the pores. The average pore size is increased and, as a result, the capillary pressure near the wellbore is made more favorable for the flow of oil.
6. Selective plugging of highly permeable zones by injecting slime-forming bacteria followed by sucrose solution which initiates the production of extracellular slimes, and aerial sweep efficiency is improved.

The target for EOR processes is the quantity of unrecoverable oil in known reservoirs and bitumen and known deposits. One of the major attributes of MEOR technologies is its low cost, but there must be the recognition that MEOR is a single process. Furthermore, reports on the deleterious activities of microorganisms in the oil field contribute to the skepticism of employing technologies using microorganisms. It is also clear that scientific knowledge of the fundamentals of microbiology must be coupled with an understanding of the geological and engineering aspects of oil production in order to develop MEOR technology.

Finally, recent developments in *upgrading* bitumen during recovery point to processes which could result in a reduction of the differential cost of bitumen upgrading (Chapter 6). These processes are based on a better understanding of the issues of asphaltene solubility effects at high temperatures, incorporation of a catalyst that is chemically precipitated internally during the upgrading, and improving hydrogen addition or carbon rejection.

4.2.8 Hybrid Processes

Hybrid approaches that involve the simultaneous use of several technologies are evolving and will see greater applications in the future. Some of the evolving options that will be tried at the field scale in the next decade include the following:

1. A mixture of steam and miscible and noncondensable hydrocarbons is being field tested as a hybrid SAGD–VAPEX approach, with apparent good success and reduction of steam–oil ratios.
2. Single horizontal laterally offset wells can be operated as moderate pressure CSS wells in combination with SAGD pairs to widen the steam chamber and reduce steam–oil ratios by about 20%.
3. Simultaneous CHOPS and SAGD can be used, with CHOPS used in offset wells until steam breakthrough occurs. Then the CHOPS wells are converted to slow gas and hot water (or steam) injection wells to control the process. The high-permeability zones generated by CHOPS should accelerate the SAGD-recovery process.
4. Incorporating PPT along with CHOPS has already been field tested with economic success, and PPT has potential applications in other hybrid approaches.
5. PPT may aid in partially stabilizing waterflood through reducing the viscous fingering and coning intensity.

In addition to hybrid approaches, the new production technologies, along with older, pressure-driven technologies, will be used in successive phases to extract more oil from reservoirs, even from reservoirs that have been abandoned after primary exploitation. Old reservoirs can be redeveloped with horizontal wells, even linking up the wells to bypassed oil because of the physics of oil film spreading between water and gas phases.

These staged approaches hold the promise of significantly increasing recoverable reserves worldwide.

4.3 ENHANCED OIL RECOVERY

EOR (*tertiary oil recovery*), also referred to as *improved oil recovery* as well as *advanced oil recovery*, is the incremental ultimate oil that can be recovered from a petroleum reservoir over oil that can be obtained by primary and secondary recovery methods (Arnarnath, 1999; Lake, 1989). EOR is often synonymous with *tertiary oil recovery* although both terms also apply to primary and secondary methods.

For tax purposes, the Internal Revenue Service of the United States has listed the projects that qualify as EOR projects (CFR, 2004, 1.43-2) and these are:

i. *Thermal Recovery Methods*
   (1) *Steam drive injection*—the continuous injection of steam into one set of wells (injection wells) to effect oil displacement toward and production from a second set of wells (production wells); (2) *cyclic steam injection*—the alternating injection of steam and production of oil with condensed steam from the same well or wells; and (3) *in situ combustion*—the combustion of oil or fuel in the reservoir sustained by injection of air, oxygen-enriched air, oxygen, or supplemental fuel supplied from the surface to displace unburned oil toward producing wells. This process may include the concurrent, alternating, or subsequent injection of water.

ii. *Gas Flood Recovery Methods*
   (1) *Miscible fluid displacement*—the injection of gas (e.g., natural gas, enriched natural gas, a liquefied petroleum slug driven by natural gas, carbon dioxide, nitrogen, or flue gas) or alcohol into the reservoir at pressure levels such that the gas or alcohol and reservoir oil are miscible; (2) *carbon dioxide–augmented waterflooding*— the injection of carbonated water, or water and carbon dioxide, to increase waterflood efficiency; (3) *immiscible carbon dioxide displacement*—the injection of carbon dioxide into an oil reservoir to effect oil displacement under conditions in which miscibility with reservoir oil is not obtained; this process may include the concurrent, alternating, or subsequent injection of water; and (4) *immiscible nonhydrocarbon gas displacement*—the injection of nonhydrocarbon gas (e.g., nitrogen) into an oil reservoir, under conditions in which miscibility with reservoir oil is not obtained, to obtain a chemical or physical reaction (other than pressure) between the oil and the injected gas or between the oil and other reservoir fluids; this process may include the concurrent, alternating, or subsequent injection of water.

iii. Chemical Flood Recovery Methods

Three EOR processes involve the use of chemicals—surfactant–polymer, polymer, and alkaline flooding (OTA, 1978).

Surfactant flooding is a multiple-slug process involving the addition of surface-active chemicals to water (Reed and Healy, 1977). These chemicals reduce the capillary forces that trap the oil in the pores of the rock. The surfactant slug displaces the majority of the oil from the reservoir volume contacted, forming a flowing oil–water bank that is propagated ahead of the surfactant slug. The principal factors that influence the surfactant slug design are interfacial properties, slug mobility in relation to the mobility of the oil–water bank, the persistence of acceptable slug properties and slug integrity in the reservoir, and cost.

Microemulsion flooding, also known as *surfactant–polymer flooding*, involves injection of a surfactant system (e.g., a surfactant, hydrocarbon, cosurfactant, electrolyte, and water) to enhance the displacement of oil toward producing wells; and (2) *caustic flooding*—the injection of water that has been made chemically basic by the addition of alkali metal hydroxides, silicates, or other chemicals.

Microemulsion flooding (*micellar/emulsion flooding*) refers to a fluid injection process in which a stable solution of oil, water, and one or more surfactants along with electrolytes of salts is injected into the formation and is displaced by a mobility buffer solution (Dreher and Gogarty, 1979; Reed and Healy, 1977). Injecting water in turn displaces the mobility buffer. Depending on the reservoir environment, a preflood may or may not be used (Venuto, 1989).

Polymer-augmented waterflooding—the injection of polymeric additives with water to improve the areal and vertical sweep efficiency of the reservoir by increasing the viscosity and decreasing the mobility of the water injected; polymer-augmented waterflooding does not include the injection of polymers for the purpose of modifying the injection profile of the wellbore or the relative permeability of various layers of the reservoir, rather than modifying the water–oil mobility ratio (Sorbie, 1991).

Alkaline flooding involves the use of aqueous solutions of certain chemicals such as sodium hydroxide, sodium silicate, and sodium carbonate that are strongly alkaline. These solutions will react with constituents present in some crude oils or present at the rock–crude oil interface to form detergent-like materials which reduce the ability of the formation to retain the oil.

Other variations on this theme include the use of steam and the means of reducing interfacial tension by the use of various solvents (Ali, 1974). The solvent approach has had some success when applied to bitumen recovery from mined tar sand but when applied to non-mined material, losses of solvent and dissolved bitumen are always an issue. However, this approach should not be rejected out of hand since a novel concept may arise that guarantees minimal (acceptable) losses of bitumen and solvent.

Miscible fluid displacement (*miscible displacement*) is an oil displacement process in which an alcohol, a refined hydrocarbon, a condensed petroleum gas, carbon dioxide, liquefied natural gas, or even exhaust gas is injected into an oil reservoir, at pressure levels such that the injected gas or alcohol and reservoir oil are miscible (Stalkup, 1983). The process may include the concurrent, alternating, or subsequent injection of water. Additionally, there can be temperature changes with corresponding physical changes as in thermal EOR and cooling during long-term waterflooding of a reservoir with cold water (Dawe, 2004).

Thermal methods for oil recovery have found most use when the oil in the reservoir has a high viscosity. For example, heavy oil is usually highly viscous (hence the use of the adjective *heavy*), with a viscosity ranging from approximately 100 cP to several million centipoises at the reservoir conditions. In addition, oil viscosity is also a function of temperature and API gravity (Speight, 2000).

Thermal EOR processes add heat to the reservoir to reduce oil viscosity and/or to vaporize the oil. In both instances, the oil is made more mobile so that it can be more effectively driven to producing wells. In addition to adding heat, these processes provide a driving force (pressure) to move oil to the producing wells.

Both forward and reverse combustion methods have been used with some degree of success when applied to tar-sand deposits. The forward combustion process has been applied to the Orinoco deposits (Terwilliger et al., 1975) and in the Kentucky sands (Terwilliger, 1975). The reverse combustion process has been applied to the Orinoco deposit (Burger, 1978) and the Athabasca deposit (Mungen and Nicholls, 1975). In tests such as these, it is essential to control

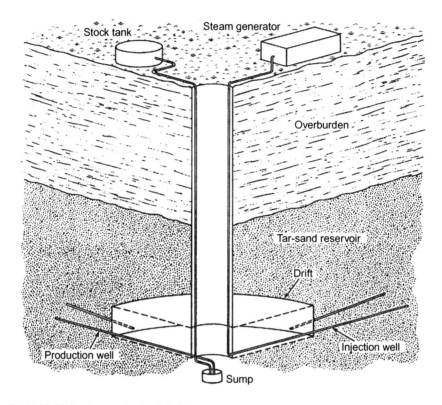

Fig. 4.1 Modified in situ extraction (Speight, 2007).

the airflow and to mitigate the potential for spontaneous ignition (Burger, 1978). A modified combustion approach has been applied to the Athabasca deposit (Mungen and Nicholls, 1975). The technique involved a heat-up phase and a production (or blowdown phase) followed by a displacement phase using a fireflood–waterflood (combination of forward combustion and waterflood, COFCAW) process.

In *modified* in situ *extraction* processes, a combination of *in situ* and mining techniques are used to access the reservoir when the organic material, such as oil sand bitumen, proves too difficult to move to the production well (Figure 4.1). A portion of the reservoir rock must be removed to enable application of the *in situ* extraction technology. The most common method is to penetrate the reservoir via a large-diameter vertical shaft, excavate horizontal drifts from the bottom of the shaft, and drill injection and production wells horizontally from the drifts. Thermal extraction processes are then applied through the wells. When the horizontal wells are drilled at or near the base of the tar-sand reservoir, the injected heat rises from the injection wells through the reservoir, and drainage of produced fluids to the production wells is assisted by gravity.

Finally, no single EOR technique is the cure-all for oil recovery. Most reservoirs are complex and the oil-reservoir system must be considered as a whole rather than as individual, but equally complex, entities.

REFERENCES

Ali, S.M. Farouq, 1974. In: Hills, L.V. (Ed.), Oil Sands Fuel of the Future Canadian Society of Petroleum Geologists, Calgary, Alberta, p. 199.

Arnarnath, A., 1999. Enhanced Oil Recovery Scoping Study. Report No. TR-113836. Electric Power Research Institute, Palo Alto, CA.

Banat, I.M., 1995. Biosurfactant production and possible uses in microbial enhanced oil recovery and oil pollution remediation. Biores. Technol. 51, 1–12.

Burger, J., 1978. In-situ recovery of oil from oil sands. In: Chilingarian, G.V., Yen, T.F. (Eds.), Developments in Petroleum Science, No. 7, Bitumens, Asphalts and Tar Sands. Elsevier, New York, NY, pp. 191–211.

Butler, R.M., Jiang, Q., 2000. Improved recovery of heavy oil by VAPEX with widely spaced horizontal injectors and producers. J. Can. Pet. Technol. 39, 48–56.

Butler, R.M., Mokrys, I.J., 1991. A new process (VAPEX) for recovering heavy oils using hot water and hydrocarbon vapor. J. Can. Pet. Technol. 30 (1), 97–106.

Butler, R.M., Mokrys, I.J., 1995a. Process and Apparatus for the Recovery of Hydrocarbons from a Hydrocarbon Deposit. United States Patent 5,407,009. April 18.

Butler, R.M., Mokrys, I.J., 1995b. Process and Apparatus for the Recovery of Hydrocarbons from a Hydrocarbon Deposit. United States Patent 5,607,016. March 4.

Bryant, R.S., Lindsey, R.P., 1996. World-wide applications of microbial technology for improving oil recovery. In: Proceedings of the SPE Symposium on Improved Oil Recovery. Society of Petroleum Engineers, Richardson TX, pp. 127–134.

Bryant, R.S., Donaldson, E.C., Yen, T.F., Chilingarian, G.V., 1989. Microbial enhanced oil recovery. In: Donaldson, E.C., Chilingarian, G.V., Yen, T.F. (Eds.), Enhanced Oil Recovery II: Processes and Operations. Elsevier, Amsterdam, the Netherlands, pp. 423–450.

CFR, 2004. Code of Federal Regulations, Internal Revenue Service. Department of the Treasury, Government of the United States, Washington DC, 1.43-2.

Chugh, S., Baker, R., Telesford, A., Zhang, E., 2000. Mainstream options for heavy oil: Part I— Cold production. J. Can. Pet. Technol. 39 (4), 31–39.

Clark, J.B., Munnecke, D.M., Jenneman, G.E., 1981. In situ microbial enhancement of oil production. Dev. Ind. Microbiol. 15, 695–701.

CRA, 2003. Pressure Pulse Technology, vol. 3. Innovative Technology Group, Conestoga-Rovers and Associates, Niagara Falls, NY, Number 1 (January). <http://www.craworld.com/en/newsevents/resources/inn_2003_jan.pdf>.

Curtis, C., Kopper, R., Decoster, E., Guzmán-Garcia, A., Huggins, C., Knauer, L., et al., 2002. Heavy-oil reservoirs. Oilfield Rev. 14 (3), 42–46.

Das, S., Butler, R.M., 1994. Effect of asphaltene deposition on the VAPEX process: A preliminary investigation using a Hele-Shaw cell. J. Can. Pet. Technol. 33 (6), 39–45.

Dawe, R.A., 2004. Miscible displacement in heterogeneous porous media. In: Proceedings of the Sixth Caribbean Congress of Fluid Dynamics, UWI, January 22–23, 2004.

Dreher, K.D., Gogarty, W.B., 1979. An overview of mobility control in micellar/polymer enhanced oil recovery processes. J. Rheol. 23 (2), 209–229.

Dunbar, R.B., 2009. Canada's Oil Sands—A World-Scale Hydrocarbon Resource. Strategy West Inc., Calgary, Alberta, Canada, <http://www.strategywest.com/downloads/StratWest_OilSands.pdf>.

Fossey, J.P., Morgan, R.J., Hayes, L.A., 1997. Development of the Pelican lake area: Reservoir considerations and horizontal technologies. J. Can. Pet. Technol. 36 (6), 53–56.

Halliburton, 2004. <http://www.onthewavefront.com/Files/Press/2004/Halliburton_19_July_04.htm>.

Jamaloei, B.Y., Dong, M., Mahinpey, N., Maini, B.B., 2012. Enhanced cyclic solvent process (ECSP) for heavy oil and bitumen recovery in thin reservoirs. Energy Fuels 26 (5), 2865–2874.

Jenneman, G.E., 1989. The potential for *in situ* microbial applications. Dev. Petrol. Sci. 22, 37–74.

Khire, J.M., Khan, M.I., 1994a. Microbially enhanced oil recovery (MEOR). Part 1. Importance and mechanisms of microbial enhanced oil recovery. Enzyme Microb. Technol. 16, 170–172.

Khire, J.M., Khan, M.I., 1994b. Microbially enhanced oil recovery (MEOR). Part 2. Microbes and the subsurface environment for microbial enhanced oil recovery. Enzyme Microb. Technol. 16, 258–259.

Krumrine, P.H., Falcone, J.S., Jr., 1987. Beyond Alkaline Flooding: Design of Complete Chemical Systems. SPE 16280, San Antonio, TX.

Lake, L.W., 1989. Enhanced Oil Recovery. Prentice-Hall, Englewood Cliffs, NJ.

Loughead, D.J., 1992. Lloydminster heavy oil production—Why so unusual? In: Ninth Annual Heavy Oil and Oil Sands Technology Symposium, Calgary, Alberta, Canada. March 11.

Luo P., Gu, Y., 2005. Effects of asphaltene content and solvent concentration on heavy oil viscosity. Paper No. SPE 97778. In: Proceedings of the SPE International Thermal Operations and Heavy Oil Symposium, Calgary, Alberta, Canada.

McCaffrey, W.J., Bowman, R.D., 1991. Recent successes in primary bitumen production. In: HOOS Technical Symposium Challenges and Innovation. Annual Technical Meeting, Petroleum Society of the Canadian Institute of Mining.

McColl, D., Mei, M., Millington, D., Kumar, C., 2008. Green Bitumen: The Role of Nuclear, Gasification, and CCS in Alberta's Oil Sands: Part I—Introduction and Overview. Study No. 119. Canadian Energy Research Institute, Calgary, Alberta, Canada.

McPhee, D., Ranger, M.J., 1998. The Geological Challenge for Development of Heavy Crude and Oil Sands of Western Canada. In: Proceedings of the Seventh UNITAR International Conference on Heavy Crude and Tar Sands, Beijing, China. October 27–30.

Meyer, R.F., Attanasi, E.D., 2003. Fact Sheet 70-03, Heavy Oil and Natural Bitumen—Strategic Petroleum Resources. United States Geological Survey, <http://pubs.usgs.gov/fs/fs070-03/>.

Meszaros, G., Chakma, A., U. of Calgary; Jha, K.N., Energy Mines and Sources Canada; Islam, M.R. 1990. Scaled model studies and numerical simulation of inert gas injection with horizontal wells. Paper 20529-MS. Proceedings. SPE Annual Technical Conference and Exhibition New Orleans, Louisiana. September 23-26.

Mitchell, D.L., Speight, J.G., 1973. The solubility of asphaltenes in hydrocarbon solvents. Fuel 52, 149.

Mok, J., 2004. <http://jonmok.com/OilSands/index.php?option=com_content&task=%20view&id=18&Itemid=44>.

Mungen, R., Nicholls, J.H., 1975. Recovery of oil from athabasca oil sands and from heavy oil deposits of northern alberta by in-situ methods. Proc. Ninth World Petroleum Congr. 6, 29.

Nenniger, J.E., Dunn, S.G., 2008. How fast is solvent-based gravity drainage. In: Proceedings of the Canadian International Petroleum Conference/SPE Gas Technology Symposium Joint Conference. 59th Annual Technical Meeting, Canadian Petroleum Society, Calgary, Alberta, Canada. June 17–19, Paper 2008-139.

OTA, 1978. Enhanced Oil Recovery Potential in the United States. Office of Technology Assessment, Washington, DC, January. NTIS order #PB-276594.

Raiders, R.A., Knapp, R.M., McInerney, M.J., 1989. Microbial selective plugging and enhanced oil recovery. J. Ind. Microbiol. 4, 215–230.

Reed, R.L., Healy, R.N., 1977. Some physico-chemical aspects of microemulsion flooding: a review. In: Shah, D.O., Schechter, R.S. (Eds.), Improved Oil Recovery by Surfactant and Polymer Flooding. Academic Press, New York, NY.

Shennan, J.L., Levi, J.D., 1987. In situ microbial enhanced oil recovery. In: Kosaric, N., Cairns, W.L., Gray, N.C.C. (Eds.), Biosurfactants and Biotechnology. Marcel Dekker, New York, NY, pp. 163–180.

Sorbie, K.S., 1991. Polymer-Improved Oil Recovery. Springer, New York, NY.

Speight, J.G., 2000. Desulfurization of Heavy Oils and Residua, second ed. Marcel Dekker, New York, NY.

Speight, J.G., 2007. Chemistry and Technology of Petroleum, fourth ed. CRC, Taylor & Francis Group, Boca Raton, FL.

Speight, J.G., 2008. Synthetic Fuels Handbook: Properties, Processes, and Performance. McGraw-Hill, New York, NY.

Speight, J.G., 2009. Enhanced Recovery Methods for Heavy Oil and Tar Sands. Gulf Publishing Company, Houston, TX.

Stalkup, F.I., 1983. Miscible Displacement. In: SPE Monograph No. 8, New York, NY.

Stosur, G.J., 1991. Unconventional EOR concepts. Crit. Rep. Appl. Chem. 33, 341–373.

Terwilliger, P.L., 1975. Fireflooding shallow tar sands. A case history. In: Proceedings of the 50th Annual Fall Meeting. Society of Petroleum Engineers, American Institute of Mechanical Engineers, Paper 5568.

Terwilliger, P.L., Clay, R.R., Wilson, L.A., Gonzalez-Gerth, E., 1975. Fireflood of the P_{2-3} sand reservoir in the Miga Field of eastern Venezuela. J. Pet. Technol. 27, 9.

Thomas, S., 2008. Enhanced oil recovery: an overview. Oil Gas Sci. Tech. 63 (1), 9–19.

Venuto, P.B., 1989. Tailoring EOR processes to geologic environments. World Oil 209 (November), 61–68.

Yazdani, A., Maini, B.B., 2008. Modeling the VAPEX process in a very large physical model. Energy Fuels 22, 535–544.

Yang, C., Gu, Y., 2005. New experimental method for measuring gas diffusivity in heavy oil by the dynamic pendant drop volume analysis (DPDVA). Ind. Eng. Chem. Res. 44 (12), 4474–4483.

CHAPTER 5

Thermal Methods of Recovery

5.1 INTRODUCTION

Bitumen (the organic component of oil sand) is often arbitrarily defined on the basis of API gravity. A more appropriate definition of bitumen, which sets it aside from heavy oil and conventional petroleum, is based on the definition offered by the US government, viz. an *extremely viscous hydrocarbon which is not recoverable in its natural state by conventional oil well production methods including currently used enhanced recovery techniques* (Chapter 1). By inference, conventional petroleum and heavy oil (recoverable by *conventional oil well production methods including currently used enhanced recovery techniques*) are different to oil sand bitumen. Be that as it may, at some stage of production, conventional petroleum (in the later stages of recovery) and heavy oil (in the earlier stages of recovery) may require the application of EOR methods for recovery (Speight, 2009).

Initially, conventional crude oil is produced from oil-bearing formations by drilling wells into a formation and recovering the oil by any one

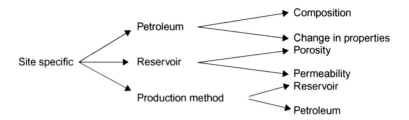

Fig. 5.1 Recovery is site specific and depends upon several variable factors (Speight, 2007).

of several possible methods (Speight, 2007, 2009). The oil is driven from the formation up through the wells (production wells) by energy stored in the formation, such as the pressure of water and dissolved natural gas (*primary recovery*). If this natural energy of the formation is expended, then energy must be injected into the formation in order to stimulate production through addition of energy to the formation through added water and gas (*secondary recovery*), followed by other more energy intensive methods of recovery (*enhanced recovery*). Furthermore, crude oil recovery depends upon several factors that, in turn, are site specific and a variety of selection criteria are involved (Figure 5.1) (Chakma et al., 1991; Islam et al., 1994; Speight, 2007, 2009).

In oil sand deposits that contain bitumen, it is necessary to initiate EOR operations as early as possible. This may mean considerable side-stepping of primary and secondary recovery operations. Thermal floods using steam and controlled *in situ* combustion methods are also used—the goal is to reduce the viscosity of the bitumen oil by heat so that it flows more easily into the production well (Pratts, 1986). Thus advanced techniques are usually variations of secondary methods with a goal of improving the *sweeping* action of the invading fluid.

The technologies applied to oil recovery involve different concepts, some of which can cause changes to the oil during production. Technologies such as alkaline flooding, microemulsion (micellar/emulsion) flooding, polymer-augmented waterflooding, and carbon dioxide miscible/immiscible flooding do not require or cause any change to the oil. The steaming technologies may cause some steam distillation that can augment the process when the steam-distilled material moves with the steam front and acts as a solvent for oil ahead of the steam front. Again, there is no change to the oil although there may be favorable compositional changes to the oil insofar as lower boiling constituents are recovered and the higher boiling constituents remain in the reservoir.

The technology where changes do occur involves combustion of the oil *in situ*. The concept of any combustion technology requires that the oil be partially combusted and that thermal decomposition occurs to other parts of the oil. This is sufficient to cause irreversible chemical and physical changes to the oil to the extent that the product is markedly different to the oil-in-place, indicating upgrading of the bitumen during the process. Recognition of this phenomenon is essential before combustion technologies are applied to oil recovery.

Furthermore, in any field where primary production is followed by a secondary or enhanced recovery method, there is the potential for noticeable differences in properties between the fluids produced. Significant differences may render the product outside of the range of acceptability for the usual refining options and force a higher demand for thermal process (i.e., coking) units.

EOR processes use *thermal*, *chemical*, or *fluid phase behavior* effects to reduce or eliminate the capillary forces that trap oil within pores, to thin the oil, or otherwise to improve its mobility or to alter the mobility of the displacing fluids (see also Chapter 4). In some cases, the effects of gravitational forces, which ordinarily cause vertical segregation of fluids of different densities, can be minimized or even used to advantage. The various processes differ considerably in complexity, in the physical mechanisms responsible for oil recovery, and in the amount of experience that has been derived from field application.

The degree to which the EOR methods will be applicable in the future will depend on the development of improved process technology. It will also depend on improved understanding of fluid chemistry, phase behavior, and physical properties, and on the accuracy of geology and reservoir engineering in characterizing the physical nature of individual reservoirs or deposits.

Variations of the EOR theme include the use of steam and solvents as the means of reducing interfacial tension. The solvent approach has had some success when applied to bitumen recovery from mined oil sand, but when applied to nonmined material, phenomenal losses of solvent and bitumen are always a major obstacle. This approach should not be rejected out of hand since a novel concept may arise that guarantees minimal (acceptable) losses of bitumen and solvent. In fact, *miscible fluid displacement* (*miscible displacement*) is a process in which an alcohol, a refined hydrocarbon, a condensed petroleum gas, carbon dioxide, liquefied natural gas, or even exhaust gas is injected into an oil reservoir, at pressure levels such that the injected gas or alcohol and reservoir oil are miscible; the process may include the concurrent, alternating, or subsequent injection of water.

The procedures for miscible displacement are the same in each case and involve the injection of a slug of solvent that is miscible with the reservoir oil followed by injection of either a liquid or a gas to sweep up any remaining solvent. As the miscible slug of solvent becomes enriched with oil as it passes through the reservoir, the composition changes, thereby reducing the effective scavenging action. However, changes in the composition of the fluid can also lead to wax deposition as well as deposition of asphaltene constituents (Speight, 2007, 2009). Therefore, caution is advised.

Microscopic observations of the leading edge of the miscible phase have shown that the displacement takes place at the boundary between the oil and the displacing phase. The small amount of oil that is bypassed is entrained and dissolved in the rest of the slug of miscible fluids; mixing and diffusion occur to permit complete recovery of the remaining oil. If a second miscible fluid is used to displace the first, another zone of displacement and mixing follows. The distance between the leading edge of the miscible slug and the bulk of pure solvent increases with the distance traveled, as mixing and reservoir heterogeneity cause the solvent to be dispersed.

Other parameters affecting the miscible displacement process are reservoir length, injection rate, porosity, and permeability of reservoir matrix, size and mobility ratio of miscible phases, gravitational effects, and chemical reactions.

Thermal recovery methods have found most use because of the high viscosity of bitumen in the deposit. For example, oil sand bitumen typically in heavy oils are highly viscous, with a viscosity ranging from several thousand centipoises to a million centipoises or more at the reservoir conditions. In addition, oil viscosity is also a function of temperature and API gravity (Speight, 2000) but the key characteristic is the immobility of the bitumen in the deposit (Chapter 1).

Thermal EOR processes (i.e., cyclic steam injection, steam flooding, and *in situ* combustion) add heat to the reservoir to reduce oil viscosity and/or to vaporize the oil. In both instances, the oil is made more mobile so that it can be more effectively driven to producing wells. In addition to adding heat, these processes provide a driving force (pressure) to move oil to producing wells. There has been some success in stem flooding the Wabasca sands (Huygen and Lowry, 1983), remembering that the Wabasca sands also contain a heavy crude oil (from the Pelican Lake oil field) that has an API of up to 21° (Chapter 4).

The potential advantages of an *in situ* process for bitumen include (1) leaving the carbon forming precursors in the ground, (2) leaving the heavy metals in the ground, (3) reducing sand handling, and (4) bringing a partially upgraded product to the surface. The extent of the upgrading can, hopefully, be adjusted by adjusting the exposure of the bitumen to the underground thermal effects.

In the *modified in situ extraction* processes, combinations of *in situ* and mining techniques are used to access the reservoir. A portion of the reservoir rock must be removed to enable application of the *in situ* extraction technology. The most common method is to enter the reservoir through a large-diameter vertical shaft, excavate horizontal drifts from the bottom of the shaft, and drill injection and production wells horizontally from the drifts. Thermal extraction processes are then applied through the wells. When the horizontal wells are drilled at or near the base of the oil sand reservoir, the injected heat rises from the injection wells through the reservoir, and drainage of produced fluids to the production wells is assisted by gravity.

Generally, as opposed to heavy oil recovery, bitumen recovery requires a higher degree of thermal stimulation because bitumen, in its immobile state, is extremely difficult (if not impossible) to move to a production well. Extreme processes are required, usually in the form of a degree of thermal conversion that produces free-flowing product oil that will flow to the well and reduce the resistance of the bitumen to flow.

Bitumen recovery processes can be conveniently divided into two categories: (1) mining methods, also called *oil mining*, and (2) nonmining methods.

In the former type of process, the oil sand must first be removed from the formation by a mining technique and then transported to a bitumen recovery center. In the latter type of process, usually (but not always correctly) termed *in situ*, bitumen (or a portion of the bitumen-in-place) is recovered from the formation by a suitable thermal method, leaving the formation somewhat less disturbed than when the mining method is employed.

5.2 NONMINING METHODS

The gravity of oil sand bitumen is usually less than 10° API depending upon the deposit, and viscosity is very high. Whereas conventional crude oils may have a viscosity of several poises (at 40°C, 105°F), the oil sand bitumen has a viscosity of the order of 50,000–1,000,000 cP or more at formation temperatures (approximately 0–10°C, 32–50°F depending upon the season). This offers a formidable (but not insurmountable) obstacle to bitumen recovery.

The successful recovery technique that is applied to one deposit/resource is not necessarily the technique that will guarantee success for another deposit. There are sufficient differences between the oil sand deposits of the United States and Canada that general applicability is not guaranteed. Hence, caution is advised when applying the knowledge gained from one resource to the issues of another resource. Although the principles may at first sight appear to be the same, the technology must be adaptable.

In principle, the *nonmining recovery of bitumen from oil sand deposits* is an enhanced recovery technique and requires the injection of a fluid into the formation through an injection wall. This leads to the *in situ* displacement of the bitumen from the recovery and bitumen production at the surface through an egress (production well). There are, however, several serious constraints that are particularly important and relate to bulk properties of the oil sand and the bitumen. In fact, both must be considered *in toto* in the context of bitumen recovery by nonmining techniques. For example, such processes need a relatively thick layer of overburden to contain the driver substance within the formation between injection and production wells.

One of the major deficiencies in applying mining techniques to bitumen recovery from oil sand deposits is (next to the immediate capital costs) the associated environmental problems. Moreover, in most of the known deposits, the vast majority of the bitumen lies in formations in which the overburden/pay zone ratio is too high. Therefore, it is not surprising that over the last two decades, a considerable number of pilot plants have been applied to the recovery of bitumen by nonmining techniques from oil sand deposits where the local terrain and character of the oil sand may not always favor a mining option.

In principle, the nonmining recovery of bitumen from oil sand deposits requires the injection of a fluid into the formation through an injection well, the *in situ* displacement of the bitumen from the reservoir, and bitumen production at the surface through an egress (production well). There are, of course, variants around this theme but the underlying principle remains the same.

There are, however, several serious constraints that are particularly important and relate to bulk properties of the oil sand and the bitumen. In fact, both must be considered in the context of bitumen recovery by nonmining techniques. For example, the Canadian deposits are unconsolidated sands with a porosity ranging up to about 45% whereas other deposits may range from predominantly low-porosity, low-permeability consolidated sand to, in a few instances, unconsolidated sands. In addition, the bitumen properties are not conducive to fluid flow under deposit conditions. Nevertheless, where the general nature of the deposits prohibits the application of a mining technique, a nonmining method may be the only feasible bitumen recovery option.

Another general constraint to bitumen recovery by nonmining methods is the relatively low injectivity of oil sand formations. Thus, it is usually necessary to inject displacement or recovery fines at a pressure such that fracturing (parting) is achieved. Such a technique therefore changes the deposit profile and introduces a series of channels through which fluids can flow from the injection well to the production well. On the other hand, the technique may be disadvantageous insofar as the fracture occurs along the path of least resistance, giving undesirable (i.e., inefficient) flow characteristics within the deposit between the injection and production wells, leaving a large part of the deposit relatively untouched by the displacement or recovery fluids.

In principle, the nonmining recovery of bitumen from oil sand deposits is an EOR technique and requires the injection of a fluid into the formation through an injection well. This leads to the *in situ* displacement of the bitumen from the deposit and bitumen production at the surface through an egress (production) well. There are, however, several serious constraints that are particularly important and relate to the bulk properties of the oil sand and the bitumen. In fact, both must be considered *in toto* in the context of bitumen recovery by nonmining techniques.

In *steam stimulation*, heat and drive energy are supplied in the form of steam injected through wells into the tar-sand formation. In most instances, the injection pressure must exceed the formation fracture pressure in order to force the steam into the oil sands and into contact with the oil. When sufficient heating has been achieved, the injection wells are closed for a soak period of variable length and then allowed to produce, first applying the pressure created by the injection and then using pumps as the wells cool and production declines.

Steam can also be injected into one or more wells with production coming from other wells (*steam drive*). This technique is very effective in heavy oil reservoirs (Speight, 2009) but has found little success during application to oil sand deposits because of the difficulty in connecting injection and production wells. However, once the flow path has been heated, the steam pressure is cycled, alternately moving steam up into the oil zone, then allowing oil to drain down into the heated flow channel to be swept to the production wells.

If the viscous bitumen in a tar-sand formation can be made mobile by admixture of either a hydrocarbon diluent or an emulsifying fluid, a relatively low-temperature secondary recovery process is possible (*emulsion steam drive*). If the formation is impermeable, communication problems exist between injection and production wells. However, it is possible to apply a solution or dilution process along a narrow fracture plane between injection and production wells.

5.2.1 Steam-Based Processes

The steam processes are the most advanced of all EOR methods in terms of field experience and thus have the least uncertainty in estimating performance, provided that a good reservoir description is available. Steam processes are most often applied in reservoirs containing heavy oil and extra heavy oil, usually in place of rather than following secondary or primary methods. Commercial application of steam processes has been in commercial practice since the early 1960s. *In situ* combustion has been field tested under a wide variety of reservoir conditions, but few projects have proven economical and advanced to commercial scale.

5.2.1.1 Steam Injection

Steam drive injection (steam injection) has been commercially applied since the early 1960s. The process occurs in two steps: (1) steam stimulation of production wells, that is, direct steam stimulation and (2) steam drive by steam injection to increase production from other wells (i.e., indirect steam stimulation).

When there is some natural reservoir energy, steam stimulation normally precedes steam drive. In steam stimulation, heat is applied to the reservoir by the injection of high-quality steam into the produce well. This cyclic process, also called *huff and puff* or *steam soak*, uses the same well for both injection and production (Speight, 2009). The period of steam injection is followed by production of reduced viscosity oil and condensed steam (water). One mechanism that aids production of the oil is the flashing of hot water (originally condensed from steam injected under high pressure) back to steam as pressure is lowered when a well is put back in production.

5.2.1.2 Cyclic Steam Stimulation

CSS (cyclic steam injection, CSI) is the alternating injection of steam and production of oil with condensed steam from the same well or wells. The process is a mature process for deep, thicker resources, such as in Cold Lake and Peace River, and involves cycling at single vertical injector/producer wells (this is sometimes referred to as *huff and puff*). An alternative incorporates steam drive between injectors and producers. While these processes originally depended on vertical wells, combinations of vertical and horizontal wells are now used.

The CSS method was first developed by Imperial Oil in the late 1950s at Cold Lake. The technology was commercialized by 1985. Initially, innovations included recycling of produced water and application of the pad drilling concept. Each pad contains a cluster of vertical and directionally drilled wells to access the bitumen-producing deposit. Drilling 20 or more wells from one pad has become commonplace. The pad design minimizes surface disturbance while directional wells provide access to a much larger area of the underground oil sand deposit than would conventional wells.

The process (*huff and puff*) system is based on producing steam in once-through steam generators or the heat recovery steam generators associated with cogeneration facilities and injecting it down the wellbore into the target formation at a temperature of about 300°C and pressures averaging 11,000 kPa. This pressure is sufficient to cause parting of the unconsolidated oil sands formation, creating paths for fluid flow. For each individual well, periods of steaming are followed by periods of soaking and then by periods of production.

CSS is a three-stage process: first, high-pressure steam is injected through a vertical wellbore for a period of time; second, the reservoir is shut in to soak; and third, the well is put into production. Typical initial cycle times for the Imperial Cold Lake development are as follows: (1) injection, 4–6 weeks; (2) soak, 4–8 weeks; and (3) production, 3–6 months.

When production rates decline, another cycle of steam injection begins. The injection–production cycle is repeated a number of times over the life of the well. The time to steam and produce the wells varies from well to well with each cycle, typically between 6 and 18 months. Expected recovery factors range between about 20% and 25% of original bitumen-in-place.

Bitumen at Cold Lake is produced from the Clearwater formation located more than 400 m below the surface. The productive zone is 100–140 ft thick. In 2006, bitumen production before royalties at Cold Lake (Imperial Oil project) exceeded 150,000 bbl per day. Dual above-ground pipelines—one delivering steam and the other carrying produced fluids back to the central processing facilities—serve multiple pads.

In addition to heating the bitumen, the high-pressure steam creates fractures in the formation, thereby improving fluid flow. High injection pressures for deep *huff and puff* processes require an overburden cover of 300 or more meters. Typical steam-to-oil ratios, the major economic factor, are 3:1–4:1. For CSS, an estimated 20–25% of the initial oil-in-place is estimated to be recoverable. This recovery rate is reportedly low when compared to SAGD methods and mining.

Although CSS is characterized by higher steam–oil ratios than with SAGD, the quality of steam used is lower and requires less energy to produce. In CSS operations, natural gas requirements can be met through produced solution gas, whereas in a SAGD operation, these amounts are comparatively minimal.

Bitumen produced by CSS tends to have a higher API gravity and is less viscous; therefore, diluent costs are reduced when compared with bitumen produced by the SAGD process. In addition, the Cold Lake region is closer to market than the Athabasca region; therefore, transportation costs are typically lower when compared with SAGD. A key focus in a CSS operation is to increase the total recovered bitumen by increasing the quantity of bitumen recovered in each cycle and/or increasing the number of cycles for which bitumen recovery is economical. The steam–oil ratio, and therefore gas costs for steam generation, is typically at its lowest point during early cycles, after which it begins to rise until the point at which bitumen production is no longer economic and the well is abandoned.

The CSS process is a well-developed process; the major limitation is that less than 30% (usually less than 20%) of the initial oil-in-place can be recovered. The process is particularly effective in reservoirs with limited vertical permeability and is best suited to operations in the Cold Lake area and the Peace River heavy oils.

5.2.1.3 Steam Drive

Steam drive involves the injection of steam through an injection well into a reservoir and the production of the mobilized bitumen and steam condensate from a production well. Steam drive is usually a logical follow-up to cyclic steam injection. Steam drive requires sufficient effective permeability (with the immobile bitumen-in-place) to allow injection of the steam at rates sufficient to raise the reservoir temperature to mobilize the bitumen.

Two expected problems inherent in the steam drive process are steam override and reservoir plugging. Any *in situ* thermal process tends to override (migrate to the top of the effected interval) because of differential density of the hot and cold fluids. These problems can be partially mitigated by rapid injection of steam at the bottom or below the target interval through a high-permeability water zone or fracture. Each of these options will raise the temperature of the entire reservoir by conduction and, to a lesser degree, by convection. The bitumen will be at least partially mobilized and the effectiveness of the following injection of steam into the target interval will be enhanced.

For a successful steam drive project, the porosity of the deposit should be at least 20%; the permeability should be at least 100 millidarcy; and the bitumen saturation should be at least 40%. The bitumen content of the deposit should be at least 800 bbl per acre-foot. The depth of the deposit should be less than 3,000 ft and the thickness should be at least 30 ft; other preferential parameters have also been noted on the basis of success with several heavy oil reservoirs. Typically, oil sand deposits do not always meet these criteria.

Other variations on this theme include the use of steam and the means of reducing interfacial tension by the use of various solvents. The *solvent extraction* approach has had some success when applied to bitumen recovery from mined oil sand, but when applied to unmined material, losses of solvent and bitumen are always a major obstacle. This approach should not be rejected out of hand since a novel concept may arise which guarantees minimal (acceptable) losses of bitumen and solvent.

5.2.1.4 Steam-Assisted Gravity Drainage

In general, the viscous nature of the bitumen and its immobility in the deposit have precluded other forms of recovery. However, bitumen recovery from deep deposits is not economical by a mining method. Therefore, the bitumen viscosity must be reduced *in situ* to increase the mobility of bitumen to flow to wellbores that bring it to the surface.

Bitumen viscosity can be reduced *in situ* by increasing deposit temperature or by injecting solvents. Steam-based thermal recovery is the primary recovery method for heavy oil (sometimes referred to as bitumen) in the Cold Lake and Peace River areas. Various steam-based methods have been shown to be inefficient for bitumen but, more recently, a method known as SAGD has been applied to the Athabasca oil sand with success.

The concept of utilizing continuous heating and production, rather than the discontinuous CSS process, led to the development of the SAGD process during the late 1970s and early 1980s. Advanced horizontal drilling technology laid the foundation for SAGD and the process is utilized in several commercial projects by various companies (Butler, 2001; Butler and Jiang, 2000; Butler and Mokrys, 1991). Provided there is sufficient permeability, the mobilized bitumen and condensed steam drain, by gravity, to the producing well, and are subsequently pumped to the surface.

SAGD makes it possible to extract bitumen from thinner reserves than with CSS, although good vertical permeability is essential. Consequently, the introduction of this technique has considerably increased the recoverable reserve category. The steam-to-oil ratio for SAGD ranges from 2.5:1, for high-quality deposits, to 3.0:1 (National Energy Board, 2004).

In the process, a pair of horizontal wells, separated vertically by about 15–20 ft, is drilled at the bottom of a thick unconsolidated sandstone deposit. Steam is injected into the upper well. The heat reduces the oil viscosity to values as low as 1–10 cP (depending on temperature and initial conditions) and develops a *steam chamber* that grows vertically and laterally. The steam and gases rise because of their low density, and the oil and condensed water are removed through the lower well. The gases produced during SAGD tend to be methane with some carbon dioxide and traces of hydrogen sulfide.

To a small degree, the noncondensable gases tend to remain high in the structure, filling the void space, and even acting as a partial insulating blanket that helps to reduce vertical heat losses as the chamber grows laterally. At the pore-size scale, and at larger scales as well, flow is through countercurrent, gravity-driven flow, and a thin and continuous oil film is sustained, giving high recoveries.

Operating the production and injection wells at approximately the same pressure as the deposit eliminates viscous fingering and coning processes, and also suppresses water influx or oil loss through permeable streaks. This keeps the steam chamber interface relatively sharp and reduces heat losses considerably. Injection pressures are much lower than the fracture gradient, which means that the chances of breaking into a thief zone, an instability problem which plagues all high-pressure steam injection processes, such as cyclic steam soak, are essentially zero.

SAGD technology offers some potential advantages over CSS in deposits that have high vertical permeability. Lower steam injection pressure generally means that SAGD can be applied to thinner deposits than with CSS, although good vertical permeability is essential. Other advantages include (1) higher recovery factors (up to 50% for the SAGD process compared to 15–20% for the CSS process), (2) lower steam–oil ratios, which reduce operating costs, and (3) use of lower pressures that allow the exploitation of shallower deposits. However, these advantages are offset to some degree by requirement of the process for injection of saturated (100% quality) steam and by much higher gas–oil ratios (associated gas production) for operations in Cold Lake compared to Athabasca.

A major advantage of the SAGD process is that an estimated 40–60% of original bitumen-in-place can be recovered compared with CSS where an estimated 20–25% of the initial oil-in-place is estimated to be recoverable.

Although commercial SAGD projects have been in operation since 2001, it is still relatively early in the development of this recovery method, and there is considerable scope for modification and improvement, in terms of energy efficiency and recovery factors.

SAGD works best in high-permeability deposits, resulting in lower injection pressures and lower steam-to-oil ratios. In a SAGD operation, several horizontal well pairs are drilled from the same pad extending as long as 1,000 m horizontally into the oil sands and about 5 m apart vertically. The top well is used to inject steam to warm up a zone around and below the injector, reducing the viscosity and mobilizing an expanding zone of bitumen, which is then produced through the lower well.

A variant of SAGD that is being tried in several projects is solvent-aided process, where injecting a combination of steam and solvent, typically butane, is expected to reduce the steam–oil ratio and accelerate the recovery rate. The recovery of most of the injected solvent is necessary for economic feasibility.

SAGD (Figure 5.2) was developed first in Canada for deposits where the immobile bitumen occurs (Dusseault et al., 1998). This process uses paired horizontal wells. Low-pressure steam continuously injected through the upper well creates a steam chamber along the walls of which the heated bitumen flows and is produced in the lower well.

Several variations of this process have been developed. One variation uses a single horizontal well, with steam injection through a central pipe and production along the annulus. Another variation involves steam injection through existing vertical wells and production through an underlying horizontal well. The key benefits of the SAGD process are

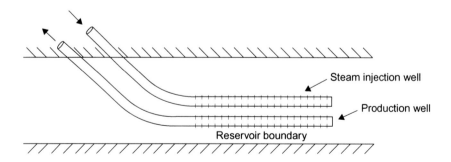

Fig. 5.2 Steam-assisted gravity drainage.

an improved steam–oil ratio and high ultimate recovery (on the order of 60–70%). The outstanding technical issues relate to low initial oil rate, artificial lifting of bitumen to the surface, horizontal well operation, and the extrapolation of the process to deposits having low permeability, low pressure, or bottom water.

To a small degree, the noncondensable gases tend to remain high in the structure, filling the void space, and even acting as a partial *insulating blanket* that helps to reduce vertical heat losses as the chamber grows laterally. At the pore scale, and at larger scales as well, flow is through countercurrent, gravity-driven flow, and a thin and continuous oil film is sustained, giving high recoveries estimated to be as large as 70–80% in suitable deposits.

Operating the production and injection wells at approximately the same pressure as the deposit adverse effects such as fingering, processes, as well as water influx or oil loss and keeps the steam chamber interface relatively sharp and reduces heat losses considerably. Injection pressures are much lower than the fracture gradient, which means that the chances of breaking into a thief zone, an instability problem which plagues all high-pressure steam injection processes, such as cyclic steam soak, are essentially zero.

Thus, the SAGD process, as for all gravity-driven processes, is extremely stable because the process zone grows only by gravity segregation, and there are no pressure-driven instabilities such as channeling, coning, and fracturing. It is vital in the SAGD process to maintain a volume balance, replacing each unit volume withdrawn with a unit volume injected, to maintain the processes in the gravity dominated domain. If bottom-water influx develops, it indicates that the pressure in the water is larger than the pressure in the steam chamber, and steps must be taken to balance the pressures. Because it is not possible to reduce the pressure in the water zone, the pressure in the steam chamber and production well region must be increased. This can be achieved by increasing the operating pressure of the steam chamber through the injection rate of steam or through reduction of the production rate from the lower well. After some time, the pressures will become more balanced and the water influx ceases. Thereafter, maintaining a volume balance carefully is necessary.

Clearly, a low-pressure gradient between the bottom water and the production well must be sustained. If pressure starts to build up in the steam chamber zone, then loss of hot water can take place as well.

In such cases, the steam chamber pressure must be reduced and per-haps also the production rate increased slightly to balance the pressures. In all these cases, the system tends to return to a stable con-figuration because of the density differences between the phases.

SAGD seems to be relatively insensitive to shale streaks and similar horizontal barriers, even up to several meters thick (3–6 ft), that other-wise would restrict vertical flow rates. This occurs because as the rock is heated, differential thermal expansion causes the shale to be placed under a tensile stress, and vertical fractures are created, which serve as conduits for steam (up) and liquids (down). As high temperatures hit the shale, the kinetic energy in the water increases, and adsorbed water on clay particles is liberated. Thus, instead of expanding thermally, dehydra-tion (loss of water) occurs and this leads to volumetric shrinkage of the shale barriers. As the shale shrinks, the lateral stress (fracture gradient) drops until the pore pressure exceeds the lateral stress, which causes verti-cal fractures to open. Thus, the combined processes of gravity segrega-tion and shale thermal fracturing make SAGD so efficient that recovery ratios of 60–70% are probably achievable even in cases where there are many thin shale streaks, although there are limits on the thickness of shale bed that can be traversed in a reasonable time.

Heat losses and deceleration of lateral growth mean that there is an economic limit to the lateral growth of the steam chamber. This limit is thought to be a chamber width of four times (4×) the vertical zone thickness. For thinner zones, horizontal well pairs would therefore have to be placed close together, increasing costs as well as providing lower total resources per well pair. In summary, the zone thickness limit (net pay thickness) must be defined for all deposits.

The cost of heat is a major economic constraint on all thermal pro-cesses. Currently, steam is generated with natural gas, and when the cost of natural gas rises, operating costs rise considerably. Thermally, SAGD is about twice as efficient as CSS, with steam–oil ratios that are now approaching two (instead of four for cyclic steam soak), for simi-lar cases. Combined with the high recovery ratios possible, SAGD will likely displace pressure-driven thermal process in all cases where the deposit is reasonably thick.

Finally, because of the lower pressures associated with SAGD, in comparison to high-pressure processes such as cyclic steam soak and steam drive, greater wellbore stability should be another asset, reducing substantially the number of sheared wells that are common in cyclic steam soak projects.

The Canadian oil sand area has several SAGD projects in progress since this region is the home of one of the largest deposits of bitumen in the world (Canada and Venezuela have the world's largest deposits). However, in spite of the success reported for the Athabasca deposit, the SAGD process is not entirely without drawbacks. The process requires amounts of make-up freshwater and large water recycling facilities as well as a high energy demand to create the steam. In addition, gravity drainage being the operative means of bitumen separation from the deposit rock, the process requires comparatively thick and homogeneous deposits; to date, the process has not been tested on a sufficiently wide variety of reservoirs or deposits to determine general applicability.

Different processes are still being developed and these include steam-assisted gas push (SAGP) and expanding-solvent SAGD (ES-SAGD) in which a hydrocarbon solvent (to facilitate complete or partial dissolution of the bitumen and help reduce its viscosity) is mixed with the steam.

There are a number of other hybrid thermal/solvent processes being tested, and a number of hybrid steam/solvent processes have been proposed which combine SAGD technology with solvent injections. The new processes are aimed primarily at improving recovery and energy efficiency and reducing water requirements.

For example, SAGP is a variation of the SAGD process involving the addition of a small amount of noncondensable gas such as natural gas or flue gas with the injected steam.

In the process, a high concentration of noncondensable gas at which the dew point is much lower than the saturation temperature of steam at a deposit pressure accumulates in the steam chamber, particularly near the top. However, noncondensed gas, mixed with a small fraction of the injected steam, flows mostly to the top through the condensation front, which can carry the pressure upward and push the oil downward (Canbolat et al., 2004). This injected gas creates a thin gas layer at the top of the deposit and provides a thermal and pressure insulation effect, which reduces the average temperature in the chamber, and the heat loss to the overburden (Butler and Yee, 2002). As a result, the steam requirement is reduced and the steam–oil ratio is improved. If the pressure in the steam chamber is higher than that of the deposit, large quantities of gas will escape from the steam chamber. However, if the operating pressure is close to or less than the surrounding pressure, gas may accumulate in the chamber reducing its temperature (the SAGP effect) and improving the steam–oil ratio (Butler, 2004).

Through improved energy efficiencies, the SAGP process presents opportunities for reducing steam consumption by up to 70% compared to SAGD.

The Fast-SAGD recovery process (Polikar et al., 2000) combines the SAGD and cyclic steam stimulation (CSS) processes in order to attempt to overcome the limitations of the SAGD process. In the FAST-SAGD process, additional single offset horizontal wells are drilled in between and parallel to SAGD well pairs. The offset wells are placed at the same elevation as the SAGD producers and can be 160–260 ft away from the SAGD well pairs. The concept relies on operating the SAGD wells until the steam chamber reaches the top of the formation and then starting a CSS operation at the offset wells at considerably higher pressure than for the SAGD wells. The purpose of injecting steam into the offset CSS well is to accelerate the growth and propagation of the steam chamber laterally. Once the interwell area between the SAGD well pairs is heated enough, ideally when the two steam chambers come into contact, the offset well is converted into a producer and the SAGD operation continues.

Other enhanced thermal processes include (1) ES-SAGD, (2) *low-pressure solvent* SAGD, (3) *tapered steam solvent* SAGD, and (4) *combustion-assisted gravity drainage* (CAGD).

The ES-SAGD process combines the benefits of steam and solvents in the recovery of bitumen (Nasr et al., 2003). This process includes higher and faster drainage rates, lower energy and water requirements, and reduced greenhouse gas emissions (GHG) emissions than current in situ thermal technologies such as cyclic steam stimulation and SAGD. Hexane is the hydrocarbon additive with the closest vaporization temperature to steam temperature and is coinjected with steam in the vapor phase, similar to the SAGD process. This dilutes the oil and in conjunction with heat reduces its viscosity around the interface of the steam chamber. Moreover, many attempts using various solvent combinations are initiated for production optimization.

The CAGD process is a novel heavy oil/bitumen recovery enhance-ment method which is a specific combination of the two *in situ* com-bustion processes (mostly in terms of the process mechanism and the combustion reactions) and the SAGD process (mostly in terms of well configuration) (Kisman and Lau, 1994; Rahnema and Mamora, 2010). This process can be substituted with other thermal heavy oil recovery enhancement methods with a potential higher recovery factor.

5.2.1.5 Nexen/OPTI Long Lake Project

The Long Lake SAGD project, a joint venture between Nexen Inc. and OPTI Canada Inc., received approval from the Boards of Directors of both companies.

The $3.4 billion SAGD project includes an integrated upgrader that uses OPTI's proprietary ORcrude™ process along with commercially available hydrocracking and gasification technology. Long Lake is the first oil sands project to integrate SAGD with an onsite upgrader. This unique configuration of proven processes is designed to significantly reduce the need to purchase outside fuel and thereby screen the project from the volatility of natural gas prices.

The project is designed to produce a premium SCO with an expected gravity of 39° API and very low sulfur content, thus yielding a high-quality refinery feedstock. The project proponents have indicated this unique configuration to result in a $5–$10 per barrel operating cost advantage over existing integrated oil sands projects.

A good example of technology integration in the oil sands is the OPTI-Nexen Long Lake project, which represents the future of Canadian oil sands expansion. This project uses SAGD technology to produce the bitumen, with the interesting feature that no natural gas is consumed to supply the high energy demand for steam injection and upgrading. Instead, the bitumen is deasphalted and the bottoms gas-ified to produce hydrogen for upgrading the deasphalted crude, and steam for SAGD production, along with power and heat sufficient for all operations.

5.2.2 Combustion Processes

In situ combustion is normally applied to reservoirs containing low-gravity oil, but has been tested over perhaps the widest spectrum of conditions of any EOR process. In the process, heat is generated within the reservoir by injecting air and burning part of the crude oil (Alderman et al., 1983; Speight, 2009). This reduces the oil viscosity and partially vaporizes the oil-in-place, and the oil is driven out of the reservoir by a combination of steam, hot water, and gas drive. Forward combustion involves movement of the hot front in the same direction as the injected air; reverse combustion involves movement of the hot front opposite to the direction of the injected air.

During *in situ combustion* or *fire flooding*, energy is generated in the formation by igniting bitumen in the formation and sustaining it in a state of combustion or partial combustion. The high temperatures generated decrease the viscosity of the oil and make it more mobile. Some cracking of the bitumen also occurs and an upgraded product rather than bitumen itself is the fluid recovered from the production wells.

The relatively small portion of the oil that remains after the displacement mechanisms have acted becomes the fuel for the *in situ* combustion process. Production is obtained from wells offsetting the injection locations. In some applications, the efficiency of the total *in situ* combustion operation can be improved by alternating water and air injection. The injected water tends to improve the utilization of heat by transferring heat from the rock behind the combustion zone to the rock immediately ahead of the combustion zone.

The performance of *in situ* combustion is predominantly determined by four factors:

1. The quantity of oil that initially resides in the rock to be burned.
2. The quantity of air required to burn the portion of the oil that fuels the process.
3. The distance to which vigorous combustion can be sustained against heat losses.
4. The mobility of the air or combustion product gases.

In many field projects, the high gas mobility has limited recovery through its adverse effect on the sweep efficiency of the burning front. Because of the density contrast between air and reservoir liquids, the burning front tends to override the reservoir liquids. To date, combustion has been most effective for the recovery of viscous oils in moderately thick reservoirs in which reservoir dip and continuity provide effective gravity drainage or operational factors permit close well spacing.

The use of combustion to stimulate oil production is regarded as attractive for deep reservoirs. In contrast to steam injection, it usually involves no loss of heat. The duration of the combustion may be less than 30 days or much as 90 days depending on requirements. In addition, backflow of the oil through the hot zone must be prevented or coking will occur.

5.2.2.1 Forward Combustion

Forward combustion involves movement of the hot front in the same direction as the injected air while reverse combustion involves movement of the hot front opposite to the direction of the injected air. In forward combustion, the hydrocarbon products released from the zone of combustion move into a relatively cold portion of the formation. Thus, there is a definite upper limit of the viscosity of the liquids that can be recovered by a forward-combustion process. On the other hand, since the air passes through the hot formation before reaching the combustion zone, burning is complete; the formation is left completely cleaned of hydrocarbons.

Forward combustion is particularly applicable to reservoirs containing mobile heavy oil or extra heavy oil with a high effective permeability. Even though lower effective reservoir permeability is required for air injection compared with steam injection, the heavy oil reservoir (or oil sand deposit) ahead of the combustion front is subject to plugging as the vaporized fluids cool and condense. Consequently, a relatively high permeability (400–1,000 millidarcy) and relatively low bitumen saturation (45–65% of pore volume) are most favorable for this process. Oil sand deposits do not usually meet these criteria.

The combustion process yields a partially upgraded product because the temperature gradient ahead of the combustion front mobilizes the lighter hydrocarbon components that move toward the cooler portion of the reservoir and mix with unheated bitumen. This mixture is eventually produced through a production well. The heavier components (e.g., coke) are left on the sand grains and are consumed as fuel for the combustion. Under certain operating conditions, a significant cost saving is attained by injecting oxygen or oxygen-enriched air rather than atmospheric air because of reduced compression costs and a lower produced gas/oil ratio.

5.2.2.2 Reverse Combustion

Reverse combustion is particularly applicable to reservoirs with lower effective permeability (in contrast to forward combustion). It is more effective because the lower permeability would cause the reservoir to be plugged by the mobilized fluids ahead of a forward-combustion front. In the reverse combustion process, the vaporized and mobilized fluids move through the heated portion of the reservoir behind the combustion front. The reverse combustion partially cracks the bitumen, consumes a portion of the bitumen as fuel, and deposits residual coke on the sand grains. In the process, part of the bitumen will be consumed as fuel and part will be deposited on the sand grains as coke leaving 40–60% recoverable. This coke deposition serves as a cementing material reducing movement and production of sand.

A modified combustion approach has been applied to the Athabasca deposit. The technique involved a heat-up phase, production (or blow-down) phase, followed by a displacement phase using a COFCAW process. In this manner, over a total 18-month period (heat-up: 8 months; blow-down: 4 months; displacement: 6 months), 29,000 bbl of upgraded oil was produced from an estimated 90,000 bbl of oil-in-place.

The addition of water or steam to an *in situ* combustion process can result in a significant increase in the overall efficiency of that process. Two major benefits may be derived. Heat transfer in the reservoir is improved because the steam and condensate have greater heat-carrying capacity than combustion gases and gaseous hydrocarbons. Sweep efficiency may also be improved because of the more favorable mobility ratio of steam bitumen compared with gas bitumen.

Modes of application include injection of alternate slugs of air (oxygen) and water or coinjection of air (oxygen) and steam. Again, the combination of air (oxygen) injection and steam or water injection increases injectivity costs that may be justified by increased bitumen recovery.

Process efficiency is affected by reservoir heterogeneity that will reduce horizontal sweep. The underburden and overburden must provide effective seals to avoid loss of injected air and produced bitumen. Process efficiency is enhanced by the presence of some interstitial water saturation. The water is vaporized by the combustion and enhances the heat transfer by convection. The combustion processes are subject to override because of differences in the densities of injected and reservoir fluids. Production wells should be monitored for, and equipped to reduce, excessively high temperatures ($>1,095°C$; $>2,000°F$) that may damage downhole production tools and tubulars.

Applying a preheating phase before the bitumen recovery phase may significantly enhance the steam or combustion extraction processes. Preheating can be particularly beneficial if the saturation of highly viscous bitumen is sufficiently great as to lower the effective permeability to the point of production being precluded by reservoir plugging. Preheating partially mobilizes the bitumen by raising its temperature and lowering its viscosity. The result is a lower required pressure to inject steam or air and move the bitumen.

Preheating may be accomplished by several methods. Conducting a reverse combustion phase in a zone of relatively high effective permeability and low bitumen saturation is one method. Steam or hot gases may he rapidly injected into a high-permeability zone in the lower portion of the reservoir. In the *fracture-assisted steam technology* (FAST) process, steam is injected rapidly into an induced horizontal fracture near the bottom of the reservoir to preheat the reservoir. This process has been applied successfully in three pilot projects in southwest Texas. Shell has accomplished the same preheating goal by injecting steam into a high-permeability bottom-water zone in the Peace River (Alberta) field. Electrical heating of the reservoir by radiofrequency waves may also be an effective method.

Using combustion to stimulate bitumen production is regarded as being attractive for deep reservoirs and, in contrast to steam injection, usually involves no loss of heat. The duration of the combustion may be short (days instead of weeks or months) depending upon requirements. In addition, backflow of the oil through the hot zone must be prevented or coking will occur. A variation of the combustion process involves use of a heat-up phase, then a blow-down (production) phase, followed by a displacement phase using a fire-water flood (COFCAW).

5.2.2.3 THAI Process

Toe-to-heel air injection (THAI) (Figure 5.3) is based on the geometry of horizontal wells that have plagued conventional *in situ* combustion. The well geometry enforces a short flow path so that any instability issues associated with conventional combustion are reduced or even eliminated.

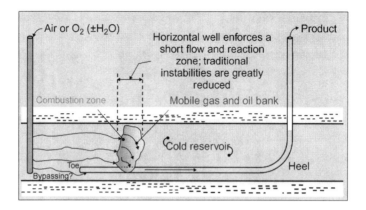

Fig. 5.3 The THAI process.

The THAI process is a method of recovery that combines a vertical air injection well with a horizontal production well. In the process, bitumen is ignited in the deposit itself, creating a vertical wall or front of burning crude (firefront) that partially upgrades the hydrocarbons in front of it and drains the crude to a producing horizontal well. By creating heat *in situ*, the process negates the need for injecting steam from the surface. The process also offers some potential for upgrading the bitumen in the deposit as the process proceeds. A THAI variant, named CAPRI, utilizes a catalyst in the horizontal well to promote the precipitation of asphaltene constituents and thus upgrade the bitumen *in situ* (Chapter 7).

In situ combustion recovery methods were tried in oil sands deposits in the 1970s and 1980s, using vertical wells, but met with little success, primarily because of an inability to control the direction of the firefront in the deposit. This generally resulted in poor production performance and often caused damage to downhole equipment. The proponents of the THAI method believe that using a horizontal production well offers better control of the firefront.

THAI is a thermal recovery process proposed for the recovery of oil sands using *in situ* combustion (Greaves and Al-Shamali, 1996; Greaves and Turta, 1997; Greaves et al., 1999, 2000). THAI combines vertical air injection wells with horizontal production wells. During this process, a combustion front is created where heavy bitumen residues left on the sand grains during the production process are burned in the deposit. This generates heat, which reduces the viscosity of bitumen, enabling it to flow, by gravity, to horizontal production wells. The combustion front sweeps the bitumen from the toe to the heel of the horizontal producing well efficiently, recovering an estimated 80% of the bitumen-in-place, while partially upgrading the bitumen *in situ*.

Other potential benefits of THAI technology include minimal natural gas and freshwater usage, partially upgraded oil quality (high-temperature oxidization of coke is left underground), lower capital and operating costs, 50% less GHG emissions, reduced diluent requirement for transportation, and the potential to operate in lower pressure, lower quality, and thinner and deeper oil sand deposits than current steam-based methods.

Whitesands Insitu Ltd. (a wholly owned subsidiary of Petrobank Energy and Resources Ltd.) is carrying out the world's first field pilot test of this technology at its Whitesands Project located at Christina Lake. The project, which was approved by the Alberta Energy Utilities Board (AEUB) in 2004, consists of three horizontal wells (1,660 ft long and 325 ft apart), three vertical air injection wells, and 19 vertical observation wells (17 for temperature and 2 for pressure observations). Positive results from the Whitesands Project have motivated the company to move to the commercial development stage with the multiphase May River Project (McColl et al., 2008). Ultimate production capacity is anticipated to be 100,000 bbl/d.

In a supplemental information package provided to the Alberta Energy and Utilities Board (AEUB) and Alberta Environment (AENV), results from an analysis of produced fluids from 3-D cell tests were reported. Oil produced using the THAI technology had an API gravity of 20.6° compared to 10° API for the product from a SAGD project. Additionally, the oil produced with THAI technology had much lower viscosity at 20°C (33 centistokes compared to 80,000 centistocks) and relatively lower levels of carbon, nitrogen, sulfur, iron, nickel, vanadium, molybdenum, saturates, resins, and asphaltene constituents.

The horizontal and vertical wells are steamed to facilitate air injection and bitumen flow. A combustion zone, with temperatures ranging between 400°C (750°F) and 700°C (1,290°F), is created when the air injected ignites the oil. The hot combustion gases coming into contact with the bitumen will cause thermal cracking and upgrading of the bitumen. The lighter oil at the combustion front and vaporized deposit water flow into the horizontal wells while the coke remains underground and functions as a fuel source for further combustion as the combustion process moves through the formation.

REFERENCES

Alderman, H., Fox, R.L., Antonation, R.G., 1983. In situ combustion pilot operations in the Wabasca heavy oil sands deposit of North Central Alberta, Canada. Paper No. 11953. Proceedings of the SPE Annual Technical Conference and Exhibition, San Francisco, CA, October 5–8.

Butler, R.M., 2001. Some recent developments in SAGD. J. Can. Petrol. Technol. 40 (1), 18–22.

Butler, R.M., 2004. The behavior of non-condensable gas in SAGD-A rationalization. J. Can. Petrol. Technol. 43 (1), 28–34.

Butler, R.M., Jiang, Q., 2000. Improved Recovery of Heavy Oil by VAPEX with Widely Spaced Horizontal Injectors and Producers. J. Can. Petrol. Technol. 39, 48–56.

Butler, R.M., Mokrys, I.J., 1991. A New Process (VAPEX) for Recovering Heavy Oils. J. Can. Petrol. Technol. 30 (1), 97–106.

Butler, R.M., Yee, C.T., 2002. Progress in the in situ recovery of heavy oils and bitumen. J. Can. Petrol. Technol. 41 (1), 31–40.

Canbolat, S., Akin, S., Polikar, M., 2004. Investigation of steam and gas push mechanism in carbonate medium. In: Proceedings of the Fifth Canadian International Symposium—55th Annual Technical Meeting, Petroleum Society of Canada, Calgary, Alberta, Canada, June 1–8, Paper 2004-223.

Chakma, A. Islam, M.R. Berruti, F. (Eds.). 1991 Enhanced Oil Recovery. AIChE Symposium Series No. 280, vol. 87. American Institute of Chemical Engineers, New York, NY, p. 147.

Dusseault, M.B., Geilikman, M.B., Spanos, T.J.T., 1998. J. Petrol. Tech. 50 (9), 92–94.

Greaves, A.T., Xia, T.X., Turta, A.T., Ayasse, C., 2000. Recent laboratory results of THAI and its comparison with other IOR process. In: Proceedings of the SPE/DOE Improved Oil Recovery Symposium held in Tulsa, Oklahoma, April 3–5, Paper SPE59334.

Greaves, M., Al-Shamali, O., 1996. In situ combustion (ISC) process using horizontal wells. J. Can. Petrol. Technol. 35 (4), 49–55.

Greaves, M., Turta, A., 1997. Oil Field In Situ Combustion Process. United States Patent No. 5,626,191, May 6.

Greaves, M., Ren, S.R., Xia, T.X., 1999. New air injection technology for IOR operations in light and heavy oil reservoirs. In: Proceedings of the SPE Asia Improved Oil Recovery Conference, Kuala Lumpur, Malaysia, October 25–26, Paper SPE57295.

Huygen, H.H.A., Lowry, W.E., Jr., 1983. Steamflooding Wabasca tar sand through the bottom-water zone—scaled model tests. SPE J. 23 (1), 92–98.

Islam, M.R., Chakma, A., Jha, K.N., 1994. Petrol. Sci. Eng. 11, 213–226.

Kisman, K.E., Lau, E.C., 1994. A new combustion process utilizing horizontal wells and gravity drainage. J. Can. Petrol. Technol. 33 (3), 31–39.

McColl, D., Mei, M., Millington, D., Kumar, C., 2008. Green Bitumen: The Role of Nuclear, Gasification, and CCS in Alberta's Oil Sands: Part I—Introduction and Overview, Study No. 119. Canadian Energy Research Institute, Calgary, Alberta, Canada.

Nasr, T.N., Beaulieu, H., Golbeck, G., Heck, G., 2003. Novel expanding solvent-SAGD process "ES-SAGD". J. Can. Pet. Technol. 42 (1), 13–16.

National Energy Board, 2004. Canada's Oil Sands: Opportunities and Challenges to 2015. National Energy Board, Calgary, Alberta, Canada.

Polikar, M., Cyr, T.J., Coates, R.M., 2000. Fast SAGD: half the wells and 30% less steam. In: Proceedings of the International Conference on Horizontal Well Technology, Calgary, Alberta, Canada, November 6–8, Paper SPE 65509.

Pratts, M., 1986. Thermal Recovery, vol. 7. Society of Petroleum Engineers, New York, NY.

Rahnema, H. and Mamora, D. 2010. Combustion-Assisted Gravity Drainage (CAGD) Appears Promising. Paper SPE 135821 presented at the 2010 Canadian Unconventional Resources & International Petroleum Conference, Calgary, Canada. October 19–21.

Speight, J.G., 2000. The Desulfurization of Heavy Oils and Residua, second ed. Marcel Dekker, New York, NY.

Speight, J.G., 2007. The Chemistry and Technology of Petroleum, fourth ed. CRC Press, Taylor and Francis Group, Boca Raton, Florida.

Speight, J.G., 2009. Enhanced Recovery Methods for Heavy Oil and Tar Sands. Gulf Publishing Company, Houston, Texas.

Upgrading During Recovery

6.1 INTRODUCTION

By all definitions, the quality of the bitumen from oil sand deposits is poor as a refinery feedstock. As in any field in which primary recovery operations are followed by secondary or enhanced recovery operations and there is a positive change in product quality, such is also the case for oil sand recovery operations. Thus, product oils recovered by thermal stimulation of oil sand deposits show some improvement in properties over those of the bitumen in place.

Bitumen upgrading is of major economic importance and oil sand deposits exist worldwide (Chapter 2). Some crude oils contain compounds such as sulfur and/or heavy metals causing additional refining problems and costs.

However, the limitations of processing oil sand bitumen depend to a large extent on the amount of nonvolatile higher molecular weight constituents (asphaltene constituents and resin constituents), which also contain the majority of the heteroatoms (i.e., nitrogen, oxygen, sulfur, and metals such as nickel and vanadium). These constituents are responsible for high yields of thermal and catalytic coke.

In catalytic processes, the asphaltene and resin constituents (as well as related constituents formed during the process) are not sufficiently mobile (i.e., they are strongly adsorbed by the catalyst) and fail to be converted to useful products either in a thermal process or in a hydroprocesses. The chemistry of the thermal reactions of some of these constituents dictates that certain reactions, once initiated, cannot be reversed and proceed to completion (Speight, 2007). Coke is the eventual product and deposits of such carbonaceous material and metal-bearing carbonaceous products deactivate the catalyst sites and eventually interfere with the catalytic process.

On this basis, *in situ* upgrading of oil sand bitumen could be a very beneficial process for leaving the unwanted elements in the deposit and increasing API gravity and overall quality of the recovered product.

Fluids produced from a well are seldom pure crude oil: in fact, a variety of materials may be produced by oil wells in addition to liquid and gaseous hydrocarbons (Speight, 2009). In addition, it has been recorded that there are noticeable differences in properties between the fluids produced (Thomas et al., 1983). The differences in elemental composition may not reflect these differences to any great extent—indeed the elemental composition of oil sand bitumen from different sources varies very little within narrow limits—and more significant differences will be evident from an inspection of the physical properties. One issue that arises from the physical property data is that such oils may be outside the range of acceptability for refining techniques other than thermal options. In addition, overloading of thermal process units will increase as the proportion of the oil sand bitumen in the refinery feedstock increases. Obviously there is a need for more and more refineries to accept larger proportions of oil sand bitumen and extra heavy oil as the refinery feedstock and have the capability to process such materials.

Technologies such as alkaline flooding, microemulsion (micellar/emulsion) flooding, polymer augmented waterflooding, and carbon dioxide miscible/immiscible flooding do not require or cause any change to the any oil sand bitumen that is recovered by these methods.

Technologies that use steam may provide some steam distillation that can augment the process when the steam distilled material moves with the steam front and acts as a solvent for oil ahead of the steam front (Jaiswal and Mamora, 2007; Northrop and Venkatesan, 1993; Sharpe and Richardson, 1995). Again, there is no chemical change to the bitumen although there may be favorable compositional changes to the bitumen insofar as lower boiling constituents are recovered and higher boiling constituents (which usually equate to coke formation during refining) remain in the deposit.

In the not too distant past, and even now, the mature and well-established processes such as visbreaking, delayed coking, fluid coking, flexicoking, propane deasphalting, and butane deasphalting were deemed adequate for upgrading heavy feedstocks.

More options are now being sought in order to increase process efficiency in terms of the yields of the desired products. Thus, it is the purpose of this chapter to (1) present an outline of the options for surface upgrading facilities and (2) to compare the concepts with *in situ* upgrading.

In view of the origin and development of new process concepts—of which the VAPEX process (Chapter 4) and the THAI process (Chapter 5) are examples—a primer on the related refinery technologies is included here.

In summary, the technologies applied to oil recovery involve different concepts, some of which can cause changes to the oil during production.

6.2 REFINERY UPGRADING

Technologies for upgrading oil sand bitumen and residua can be broadly divided into *carbon rejection* and *hydrogen addition* processes (Speight, 2007, 2011; Speight and Ozum, 2002).

Carbon rejection processes are those processes in which hydrogen is redistributed among the various components, resulting in fractions with increased hydrogen/carbon atomic ratios (distillates) and fractions with lower hydrogen/carbon atomic ratios (coke). On the other hand, *hydrogen addition* processes involve reacting heavy crude oils with an external source of hydrogen and result in an overall increase in hydrogen/carbon ratio. Finally, *separation processes* can be used to remove the difficult to process fractions that hinder and have a detrimental effect on catalytic processes (Speight, 2007; Speight and Ozum, 2002).

New bitumen processing capacity needs to be added to existing refineries, or it could be built into separate, stand-alone upgrading facilities. If the bitumen is too viscous to transport by pipeline (and it is unless diluted by a solvent), and/or there is the need for heat or energy at the production site, oil sand bitumen upgrading in the field is attractive and may avoid extensive modifications of existing refineries. Traditional processes such as coking or hydrocracking are very expensive processes and typically require a large scale to be viable. Thus, petroleum refining is now in a significant transition period as the industry moves further into the twenty-first century, and the demand for petroleum and petroleum products has shown a sharp growth in recent decades (Speight, 2011).

In order to satisfy the changing patterns of feedstock slate and product demand, significant investments in refining conversion processes will be necessary to profitably utilize heavy feedstocks. The most efficient and economical solution to this problem will depend to a large extent on individual refinery situations and this may require the use of two or more technologies in series rather than an attempt to develop a new one-stop technology for bitumen conversion (Speight, 2011).

The manner in which refineries convert oil sand bitumen into low boiling high-value products has become a major focus of operations with new concepts evolving into new processes (Khan and Patmore, 1998; Speight, 2000, 2007). Even though they may not be classed as conversion processes *per se*, pretreatment processes for removing asphaltene constituents, metals, sulfur, and nitrogen constituents are also important and can play an important role.

For example, refineries can have any one, or a conjunction, of several configurations, but the refinery of the future will, of necessity, be required to be a *conversion refinery*, which incorporates all the basic building blocks found in both the *topping refinery* and the *hydroskimming refinery*, but it also features gas oil conversion plants such as catalytic cracking and hydrocracking units, olefin conversion plants such as alkylation or polymerization units, and, frequently, coking units for sharply reducing or eliminating the production of residual fuels. Many such refineries also incorporate solvent extraction processes for manufacturing lubricants and petrochemical units with which to recover high-purity petrochemical feedstocks and petrochemical products (Speight, 2007).

New processes for the conversion of oil sand bitumen will probably be used perhaps not in place of but in conjunction with visbreaking and coking options with some degree of hydroprocessing as a primary conversion step (Dickenson et al., 1997). In addition, other processes may replace or, more likely, augment the deasphalting units in many refineries. An exception, which may become the rule, is the upgrading of bitumen from oil sands (Speight, 1990, 2000, 2007, 2011). The bitumen is subjected to either delayed coking or fluid coking as the *primary upgrading* step without prior distillation or topping. After primary upgrading, the product streams are hydrotreated and combined to form an SCO that is shipped to a conventional refinery for further processing.

Conceivably, oil sand bitumen could be upgraded in the same manner and, depending upon the upgrading facility, upgraded further for marketing. However, it is not to be construed that bitumen upgrading will *always* involve a coking step as the primary upgrading step. Other options, including some presented elsewhere (Speight, 2007), could well become predominant methods for upgrading in the future.

Thermal cracking processes offer attractive methods of bitumen conversion at low operating pressure without requiring expensive catalysts. Currently, the most widely operated residuum conversion processes are visbreaking and delayed coking, and other processes which have also received some attention for bitumen upgrading include partial upgrading (a form of thermal deasphalting), flexicoking, the Eureka process, and various hydrocracking processes.

A *partial coking* or *thermal deasphalting* process provides an upgrading of bitumen to a low level. In partial coking, the hot water process froth is distilled at atmospheric pressure, and minerals and water are removed. A dehydrated mineral-free bitumen product is obtained that contains most of the asphaltene constituents and coke precursors. Thermal cracking begins as the liquid temperature passes 340°C (645°F)—well above the range of the steam-based recovery processes. The distillation is continued into the range 370–450°C (700–840°F). With slow heating (10°C, 18°F, temperature rise per hour), the coke production rate is approximately 1% w/w of feed per hour. As the coke forms about the entrained mineral particles, 1–4% w/w coke up to 50% v/v of the feed is recovered as distillate.

Solvent deasphalting processes have not realized their maximum potential in terms of use with oil sand bitumen. With ongoing improvements in energy efficiency, such processes would display their effects in combination with other processes. Solvent deasphalting allows removal of sulfur and nitrogen compounds as well as metallic constituents by balancing yield with the desired feedstock properties (Ancheyta and Speight, 2007; Ditman, 1973; Gearhart, 1980; Low et al., 1995; Mitchell and Speight, 1973; Northrup and Sloan, 1996; RAROP, 1991, pp. 9, 91, 95, 97; Speight and Ozum, 2002).

The *solvent deasphalting process* is a mature process (Speight, 2007; Speight and Ozum, 2002), but there are new options that also provide for deasphalting various feedstocks (Speight, 2011). In the process, the feedstock is mixed with dilution solvent from the solvent accumulator and then cooled to the desired temperature before entering the extraction tower. Because of its high viscosity, the charge oil can neither be cooled easily to the required temperature nor will it mix readily with solvent in the extraction tower. By adding a relatively small portion of solvent upstream of the charge cooler (insufficient to cause phase separation), the viscosity problem is avoided.

The yield of deasphalted oil varies with the feedstock, but the deasphalted oil does make less coke and more distillate than the feedstock. The metals content of the deasphalted oil and the nitrogen and sulfur contents in the deasphalted oil are also related to the yield of deasphalted oil. However, the process parameters for a deasphalting unit must be selected with care according to the nature of the feedstock and the desired final products.

In the first case, the *choice of solvent* is vital to the flexibility and performance of the unit. The solvent must be suitable, not only for the extraction of the desired oil fraction but also for control of the yield and/or quality of the deasphalted oil at temperatures which are within the operating limits.

The main consideration in the selection of the *operating temperature* is its effect on the yield of deasphalted oil. If the temperature is too high (i.e., close to the critical temperature of the solvent), the operation becomes unreliable in terms of product yields and character. If the temperature is too low, the feedstock may be too viscous and have an adverse effect on the contact with the solvent in the tower. Liquid propane is by far the most selective solvent among the light hydrocarbons used for deasphalting—*iso*-butane and *n*-butane are also used.

At temperatures ranging from 38°C to 65°C (100–150°F), most hydrocarbons are soluble in propane while asphaltene constituents and resin constituent compounds are not, thereby allowing rejection of these compounds resulting in a drastic reduction (relative to the feedstock) of the nitrogen content and the metals in the deasphalted oil.

Solvent composition is an important variable for the deasphalting process. The use of a single solvent may (depending on the nature of the solvent) limit the range of feedstocks that can be processed in the deasphalting process. When a deasphalting unit is required to handle a variety of feedstocks and/or produce various yields of deasphalted oil (as is the case in these days of variable feedstock quality), a dual solvent may be the only option to provide the desired flexibility. For example, a mixture of propane and *n*-butane might be suitable for feedstocks that vary from vacuum residua to heavy gas oils that contain asphaltic materials. Adjusting the solvent composition allows the most desirable product quantity and quality within the range of temperature control.

Besides the solvent composition, the *solvent/oil ratio* also plays an important role in a deasphalting operation. The ratios of propane/oil required vary from 6 to 1 to 10 to 1 by volume, with the ratio occasionally being as high as 13 to −1. Since the critical temperature of propane is 97°C (206°F), this limits the extraction temperature to about 82°C (180°F). Therefore, propane alone may not be suitable for high viscosity feedstocks because of the relatively low operating temperature.

Although *n*-pentane is less selective for metals and carbon residue removal, it can increase the yield of deasphalted oil from a heavy feed by a factor of 2 to 3 over propane (Speight, 2000, 2007). However, if the content of the metals and carbon residue of the pentane-deasphalted oil is too high (defined by the ensuing process), the deasphalted oil may be unsuitable as a cracking feedstock. In certain cases, the nature of the cracking catalyst may dictate that the pentane-deasphalted oil be blended with vacuum gas oil that, after further treatment such as hydrodesulfurization, produces a good cracking feedstock.

6.3 *IN SITU* UPGRADING

The potential advantages of an *in situ* process for bitumen upgrading include (1) leaving the carbon forming precursors in the ground, (2) leaving the heavy metals in the ground, (3) reducing sand handling, and (4) bringing a partially upgraded product to the surface. The extent of the upgrading can, hopefully, be adjusted by adjusting the exposure of the bitumen of oil sand to the underground thermal effects.

In situ conversion, or underground refining, is a promising new technology to tap the extensive deposits of oil sand bitumen. The new technology (United States Patent 6,016,867; United States Patent 6,016,868) features the injection of high-temperature, high-quality steam and hot hydrogen into a formation containing bitumen to initiate conversion of the higher boiling constituents into lower boiling products (Ovalles et al., 2001). In effect, the bitumen undergoes partial underground refining that converts it into a synthetic crude oil (SCO) (or *syncrude*). The heavier portion of the syncrude is treated to provide the fuel and hydrogen required by the process, and the lighter portion is marketed as a conventional crude oil.

Thus, below ground, superheated steam and hot hydrogen are injected into an oil sand bitumen or bitumen formation, which simultaneously produces the oil sand bitumen or bitumen and converts it *in situ* (i.e., within the formation) into syncrude. Above ground, the heavier fraction of the syncrude is separated and treated on-site to produce the fuel and hydrogen required by the process, while the lighter fraction is sent to a conventional refinery to be made into petroleum products (United States Patent 6,016,867; United States Patent 6,016,868).

A gap is a technology for partial upgrading. Hydrogen addition must be used during upgrading in order to stabilize the upgraded bitumen. This means that the cost of partial upgrading is not much reduced as compared to full upgrading. Therefore, the only choice currently is no upgrading or full upgrading. Other goals could be to achieve breakthroughs in upgrading technologies such as nonthermal coking methods that would use far less energy, or gasification at 800°C, which is far lower than current commercial temperatures. The technology where changes do occur involves combustion of the oil *in situ*. The concept of any combustion technology requires that the oil be partially combusted and that thermal decomposition take place in other parts of the oil. This is sufficient to cause irreversible chemical and physical changes to the oil to the extent that the product is markedly different to the oil in place. Recognition of this phenomenon is essential before combustion technologies are applied to oil recovery.

Although this improvement in properties may not appear to be too drastic, nevertheless it usually is sufficient to provide major advantages for refinery operators. Any incremental increase in the units of hydrogen/carbon ratio can save significant amounts of costly hydrogen during upgrading. The same principles are also operative for reductions in the nitrogen, sulfur, and oxygen contents. This latter feature also improves catalyst life and activity as well as reduces the metals content.

In short, *in situ* recovery processes (although less efficient in terms of bitumen recovery relative to mining operations) may have the added benefit of *leaving* some of the more obnoxious constituents (from the processing objective) in the ground. Processes that offer the potential for partial upgrading during recovery are varied but usually follow from a surface process. Not that this be construed as an easy task: there are many disadvantages that arise from attempting *in situ* upgrading. Nevertheless the concepts are the following.

6.3.1 Partial Combustion or Gasification
The mobilization of oil sand bitumen in the deposit by partial combustion is not a new idea and still has many hurdles to overcome before it can be considered close to commercial. However, the product oil is likely less viscous.

6.3.2 Solvent Recovery
The application of light hydrocarbon solvents to reduce or eliminate natural gas for steam generation has received significant recent interest. These light hydrocarbons also have a natural tendency to cause asphaltene constituents (and, depending on the solvent, resin constituents) to separate, thereby offering promise of some *in situ* upgrading. VAPEX is the most advanced process in this area.

6.3.3 Combined Thermal-Solvent Recovery
An extension of solvent recovery is the combined use of solvents and thermal stimulation to achieve some degree of *in situ* upgrading. Several companies or joint ventures are known to be piloting variations of this approach. In all cases, a minor factor is a degree of upgrading that may occur in new recovery methods but a potentially major factor is the likely conversion of bitumen-based residues in the future for energy, power, and hydrogen at production or upgrading stages and the

possible application of mild *in situ* field upgrading to reduce dependence on diluent for transport to distant refineries.

The potential move to *less severe* primary upgrading will place more emphasis on *conversion* at the secondary stage as well as heteroatom removal. The desire to reduce overall hydrogen consumption will place emphasis on lower light by-product production and targeted hydrogen addition to SCO cuts.

6.3.4 Hydroretorting
The benefits of introduction of hydrogen during *in situ* retorting offer much promise. The possible application of such methods for selective separation of the metal constituents is an obvious benefit. For example, partial oxidation in the presence of steam may produce hydrogen for immediate pickup and result in integrated recovery and significant upgrading.

6.3.5 Selective Separations
Physical and chemical separations into fractions might lead to segregated and more targeted process steps, including more efficiently targeted hydrogen addition. There may be some overlap here with demetallization.

6.3.6 Bulk Thermal Processes
Visbreaking and variants, and the recently demonstrated ORMAT process are examples of bulk thermal processes that convert residues without progressing all the way to solid coke. These processes have significant potential integrated with deasphalting to produce residues of varying yields on bitumen to meet future alternative energy and hydrogen production needs.

In situ combustion has long been used as an enhanced oil recovery method and the potential for upgrading the oil during the process has long been recognized (Castanier and Brigham, 2003). For oil sand bitumen, numerous field observations have shown upgrading of 2–6° API for oil sand bitumen undergoing combustion (Lim et al., 2010; Ramey et al., 1992). During *in situ* combustion of oil sand bitumen, temperatures of up to 700°C can be observed at the combustion front.

In situ combustion is injection of an oxidizing gas (air or oxygen enriched air) to generate heat by burning a portion of the oil. Most of the oil is driven toward the producers by a combination of gas drive (from the combustion gases) and steam and water drive. This process is also called fire flooding to describe the movement of the burning front inside the deposit. Based on the respective directions of front propagation and air flow, the process can be forward, when the combustion front advances in the same direction as the air flow, or reverse, when the front moves against the air flow.

Forward combustion can be further characterized as *dry* when only air or enriched air is injected or *wet* when air and water are coinjected. In the process, air is injected in the target formation for a short time, usually a few days to a few weeks and the oil in the formation is ignited. Ignition can be induced using downhole gas burners, electrical heaters, and/or injection of pyrophoric agents (not recommended) or steam. In some cases, auto-ignition occurs when the deposit temperature is fairly high and the oil reasonably reactive. This often happens for California heavy oil and extra heavy oil.

After ignition, the combustion front is propagated by a continuous flow of air. As the front progresses into the deposit, several zones can be found between the injector and the producer as a result of heat, mass transport, and the chemical reactions occurring in the process. The burned zone is the volume already burned. This zone is filled with air and may contain small amounts of residual unburned organic solids. As it has been subjected to high temperatures, mineral alterations are possible. Because of the continuous air flow from the injector to the burned zone, temperature increases from the temperature of the injected air at the injector to near combustion front temperature near the combustion front. There is no oil left in this zone.

The combustion front is the highest temperature zone. It is very thin, often no more than several inches thick. It is in that region that oxygen combines with the fuel and high-temperature oxidation occurs. The products of the burning reactions are water and carbon oxides. The fuel is often misnamed coke. In fact, it is not pure carbon but a hydrocarbon with H/C atomic ratios ranging from about 1 to 2.0. This fuel is formed in the thermal cracking zone just ahead of the front and is the product of cracking and pyrolysis, which is deposited on the rock matrix. The amount of fuel burned is an important parameter because it determines how much air must be injected to burn a certain volume of deposit.

Chemical reactions are of two main categories: oxidation, which occurs in the presence of oxygen, and (2) pyrolysis, which is caused mainly by elevated temperatures.

In general at low temperature, oxygen combines with the oil to form oxidized hydrocarbons such as peroxides, alcohols, or ketones. This generally increases the oil viscosity but could increase oil reactivity at higher temperature. When oxygen contacts the oil at higher temperature, combustion occurs resulting in production of water and carbon oxides.

Of all the reactions that can occur during *in situ* combustion, only low-temperature oxidation can increase the viscosity of the oil. If the fireflood is conducted properly, low-temperature oxidations are minimized because most of the oxygen injected is consumed at the burning front.

Distillation allows transport and production of the light fractions of the oil leaving behind the higher boiling material—as in refinery distillation (Speight, 2007)—which often contains the majority of the undesirable compounds; the latter may contain sulfur or metals.

Forward *in situ* combustion by itself is already an effective *in situ* upgrading method. Improvements to the product by as much as 6° API have been observed (Ramey et al., 1992).

Another possible *in situ* upgrading technique involves a combination of solvent injection and combustion. Cyclic oil recovery has numerous advantages both technically and economically. It can also be easily optimized in a specific reservoir or deposit. Cyclic injection of solvents, either gas or liquid, followed by *in situ* combustion of a small part of the deposit can be employed to increase the temperature near the well but also to clean the wellbore region of all the residues left by the solvents. Alternate slugs of solvent and air would be injected and production would occur after each solvent slug injection and after each combustion period. The process could be repeated until an economic limit is reached. One important fact to note is that both solvent injection and *in situ* combustion have been proven to be effective in a variety of reservoirs; however, the combination of the two methods has never been tried.

The most significant effect will be the precipitation and/or deposition of asphaltene constituents or resin constituent. The product is expected to be slightly upgraded by the solvent cycle. Unlike the classic well to well *in situ* combustion, we would only try to improve near well-bore conditions by burning the solid residues left after the solvent cycle. The benefits of using combustion at this stage are expected to include (1) productivity improvement through removal of the heavy ends left from the solvent cycle, (2) possible deactivation of the clays near the wellbore due to the high temperature of the combustion, and (3) reduced viscosity of the oil due to temperature increase.

Downhole upgrading of virgin Athabasca oil sand bitumen has been investigated in a series of experiments using the THAI process, which uses combinations of vertical injection wells and horizontal producer wells, arranged in a direct or staggered line drive (Greaves and Xia, 2004; Xia and Greaves, 2006). Downhole upgrading of the bitumen was significant, with the API gravity of the produced oil increasing by an average of 8° API, compared to the original bitumen. The produced oil viscosity was also dramatically reduced and SARA (*s*aturates, *a*ro-matics, *r*esins, *a*sphaltene) analysis was used to assess the quality of the produced oil, showing that the saturate fraction of the bitumen was increased from approximately 16% by weight to 72% by weight.

The THAI process could well have a wider range of application than SAGD, but in any case, a detailed knowledge of the reservoir or deposit is essential. SAGD generally works best in relatively thick (40 m) homogeneous pay zones. It is possible that the THAI process will be effective down to about 6 m thickness, as is common in many Saskatchewan oil sand bitumen pools.

The CAPRI process involves the addition of gravel-packed cata-lyst, as used in a conventional refinery, between the tubing and the horizontal wellbore. Test results have shown the technique to add 6–8 API points on top of the THAI *in situ* upgrades. Based on these data, the combination could deliver *in situ* upgrading to above the 22° API requirement for produced fluids that can be transported by pipe-line without diluent, which also represents a major saving in surface upgrading and refining costs.

6.4 FIELD UPGRADING

In the last few years, some producers have evaluated field upgrading (moderate upgrading) processes. Interest therein was driven by concern for the future supply of diluent (typically gas field condensate) used to reduce bitumen viscosity for pipeline transport to distant refineries. The issue may resolve itself in time: it is now anticipated that more bitumen will be fully upgraded, and bitumen might increasingly be transported with synthetic crude. Consequently, there are currently no projects that plan to employ field upgrading for transport purposes.

The kinds of processes that have been proposed for field upgrading typically involve solvent-based deasphalting or mild, precoking thermal processes, such as visbreaking.

In reviewing, some of the challenges face major development of the oil sands industry, and some themes emerge that point to future "greenfield" plants. The upgrader of the future will capitalize on, or address, the following trends: (1) taking advantage of some relatively minor upgrading at the recovery stage, (2) taking advantage of the necessity to move to alternative energy and hydrogen sources, particularly internally generated residues—a trend with very large impact on main upgrader plant process selection, (3) addressing major environmental concerns in an integrated way, and (4) meeting future crude quality trends in current planning, not in a reactive way with retrofits.

The major focus of this study is the identification of technologies that are directed to transportation fuels via refineries, and addressing the niche opportunities afforded by such as the petrochemicals industry.

The future development of the industry will also need to address the eventual, but not sudden, adoption of alternative energies such as that for fuel cells, and even hydrogen. Many of the technologies that will meet some of those long-term needs are included in the review. They include hydrocracking, gasification, and syngas conversion. So, while the bitumen upgrading step has seen a major cost in competing with, or replacing declining conventional crudes, the very need for upgrading may lead to a resource more versatile in addressing long-term, radical changes in the energy economy.

REFERENCES

Ancheyta, J., Speight, J.G., 2007. Hydroprocessing of Heavy Oils and Residua. CRC-Taylor and Francis Group, Boca Raton, FL.

Castanier, L.M., Brigham, W.E., 2003. Upgrading of crude oil via *in situ* combustion. J. Petroleum Sci. Eng. 39 (1-2), 125–136.

Dickenson, R.L., Biasca, F.E., Schulman, B.L., Johnson, H.E., 1997. Refiner options for converting and utilizing heavy fuel oil. Hydrocarbon Process. 76 (2), 57.

Ditman, J.G., 1973. Deasphalt to get feed for lubes. Hydrocarbon Process. 52 (5), 110–113.

Gearhart, J.A., 1980. Solvent treat resids. Hydrocarbon Process. 59 (5), 150–151.

Greaves, Xia, T.X., 2004. Downhole Upgrading of Wolf Lake Oil Using THAI/CAPRI Processes—Tracer Tests. Preprints. Division of Fuel Chemistry, American Chemical Society, 49 (1), 69–72.

Jaiswal, N.J., Mamora, D.D., 2007. Distillation effects in heavy-oil recovery under steam injection with hydrocarbon additives. Paper No. 110712. In: Proceedings of the SPE Annual Technical Conference and Exhibition, Anaheim, CA, November 11–14.

Khan, M.R., Patmore, D.J., 1998. In: Speight, J.G. (Ed.), Petroleum Chemistry and Refining. Taylor & Francis, Washington, DC (Chapter 6).

Lim, G., Ivory, J., Coates, R., 2010. System and Method for the Recovery of Hydrocarbons by In Situ Combustion. United States Patent 7,740.062, June 22.

Low, J.Y., Hood, R.L., Lynch, K.Z., 1995. Preprints Div. Petrol. Chem. Am. Chem. Soc. 40, 780.

Mitchell, D.L., Speight, J.G., 1973. The solubility of asphaltenes in hydrocarbon solvents. Fuel 52, 149.

Northrop, P.S., Venkatesan, V.N., 1993. Analytical steam distillation model for thermal enhanced oil recovery processes. Ind. Eng. Chem. Res. 32, 2039–2046.

Northrup, A.H., Sloan, H.D., 1996. Annual Meeting. National Petroleum Refiners Association. Houston, TX, Paper AM-96-55.

Ovalles, C., Vallejos, C., Vasquez, T., Martinis, J., Perez-Perez, A., Cotte, E., et al., 2001. Extra-heavy crude oil downhole upgrading process using hydrogen donors under steam injection conditions. Paper No. 69692. In: Proceedings of the SPE International Thermal Operations and Heavy Oil Symposium. Porlamar, Margarita Island, Venezuela, March 12–14.

Ramey, H.J., Jr., Stamp, V.V., Pebdani, F.N., 1992. Case history of South Belridge, California. SPE Paper No. 24200. In: Proceedings of the Ninth SPE/DOE EOR Symposium. In Situ Combustion Oil Recovery, Tulsa, Oklahoma, April 21–24.

RAROP, 1991. RAROP Oil Sand Bitumen Processing Handbook. Research Association for Residual Oil Processing. In: Noguchi, T. (Chairman). Ministry of Trade and International Industry (MITI), Tokyo, Japan.

Sharpe, H.N., Richardson, W.C., 1995. Representation of steam distillation and in situ upgrading processes in a heavy oil simulation. Paper No. 30301. In: Proceedings of the SPE International Heavy Oil Symposium, Calgary, Alberta, Canada, June 19–21.

Speight, J.G., 1990. Tar Sand. In: Speight, J.G. (Ed.), Fuel Science and Technology Handbook. Marcel Dekker, New York, NY (Chapters 12–16).

Speight, J.G., 2000. The Desulfurization of Oil Sand Bitumen and Residua, second ed. Marcel Dekker, New York, NY.

Speight, J.G., 2007. The Chemistry and Technology of Petroleum, fourth ed. CRC, Taylor & Francis Group, Boca Raton, FL.

Speight, J.G., 2009. Enhanced Recovery Methods for Heavy Oil and Tar Sands. Gulf Publishing Company, Houston, TX.

Speight, J.G., 2011. The Refinery of the Future. Gulf Professional Publishing, Elsevier, Oxford.

Speight, J.G., Ozum, B., 2002. Petroleum Refining Processes. Marcel Dekker, New York, NY.

Thomas, K.P., Barbour, R.V., Branthaver, J.F., Dorrence, S.M., 1983. Composition of oils produced during an echoing in situ. Combustion Utah Tar Sand. Fuel 62, 438–444.

Xia, T.X., Greaves, M., 2006. In situ upgrading of Athabasca Tar Sand bitumen using THAI. Chem. Eng. Res. Design 84, 856–864.

Environmental Impact

7.1 INTRODUCTION

As Canadian oil sands production is set to enter a period of strong growth and expansion, a number of environmental issues and challenges are facing the industry. Most attention has been given to accelerating greenhouse gas emissions, but other environmental issues such as surface disturbance and water conservation also represent serious problems for the operators of oil sand projects and need to be weighed against the economic aspects of oil sand development (Charpentier et al., 2009; NEB, 2006; Swart and Weaver, 2012).

In the perspective of peak oil, Canada's huge reserves of unconventional oil have the world's attention. It is often claimed that nonconventional oil production such as oil sands production may bridge the coming gap between the world's soaring oil demand and global oil supply.

The world's nonconventional oil initially in place could amount to as much as 7 trillion barrels (7×10^{12} bbl). Oil sand deposits in Canada and the United States as well as extra heavy oil in Venezuela account for the majority of these resources. However, the amount of bitumen (and, hence, synthetic crude oil) that could be recovered from these resources is very uncertain.

The strong growth in oil demand indicates that Canada's vast resources of oil sand may have a market. However, as the oil sand industry strives to exploit these resources, significant challenges must be overcome, most importantly higher natural gas prices, capital cost over-runs, and environmental impacts.

Critics contend that government and industry measures taken to minimize environmental and health risks posed by large-scale mining operations are inadequate, causing damage to the natural environment. In fact, there are those critics who would have oil sand development stopped—there appears to be concern that oil sand development rapes the environment and will leave it a disaster area for future generations. This is not quite the case.

It has long been recognized that there is the need for responsible resource development, and the various levels of government have put the criteria in place to assure minimal environmental impact through (1) science-based precautionary limits that tell us when ecosystems are threatened and (2) improvement of the systems and approaches for monitoring and addressing the impacts of oil sand development on the climate, air, freshwater, boreal forest, and wildlife. In fact, the establishment and implementation of an effective oil sands monitoring is fundamental to the long-term environmental sustainability and economic viability of a rapidly growing oil sands industry in Canada (Dowdeswell et al., 2010) or, for that matter, in any country that seeks to follow development of indigenous oil sand resources.

7.2 SURFACE DISTURBANCE

The surface disturbance from mining operations and processing of bitumen includes land clearing, and disturbance of surface strata and soil. These activities result in deforestation of forests and woodlands, and have a negative impact on fish and wildlife populations.

The open-pit mining of the Athabasca oils sand deposits destroys the boreal forest and muskeg, as well as bringing about changes to the natural landscape. The Alberta government does not require companies to restore the land to *original condition* but to *equivalent land capability*. This means that the ability of the land to support various land uses after reclamation is similar to what existed, but that the individual land uses will not necessarily be identical. Since the government considers agricultural land to be equivalent to forest land, mined land is being reclaimed to use as pasture for buffalo, rather than restoring it to the original boreal forest and muskeg.

Another (major) issue is the development of methods that will reduce the land required for out-of-pit overburden dumps, open-pit operations, and tailings management areas (ERCB, 2008; Mikula et al., 2008). Current industry practice is to leave large areas of land to remain in a disturbed state over many years during which natural processes work to reestablish the landscape. Oil sands operators are working to preserve an environment that has an ecological capability at least equal to its condition before the start of oil sands operations. The *in situ* process is much less harmful in terms of surface damage and results in limited negative environmental impact on forests, wildlife, and fisheries.

7.3 WATER

Bitumen recovery by mining and *in situ* operations consumes large volumes of water. Water requirements for oil sands projects range from 2.5 units to 4.0 bbl of water for each barrel of bitumen produced.

The primary challenge for process water is that no large-scale water treatment facilities exist near the oil sands—as a result, process water must be recycled. Groundwater aquifers are used as the source of process water and a typical operating procedure is the disposal of process-affected water to deep aquifers. The decision to use groundwater or surface water is dependent on whether a source of surface water is available or if it is necessary to drill a well to access subsurface aquifers. Developers have also been devising methods of using brackish water from underground aquifers (NEB, 2006).

In 2003, Alberta's Environment Minister initiated a committee to find ways to reduce the oil and gas industry's consumption of freshwater. As part of the province's long-term water strategy, limits may be placed on the volume of freshwater that companies are allowed to use. Alberta Environment Department does not require a license for withdrawal of saline groundwater. There are also industrial trends to use brackish water in place of freshwater (NEB, 2006).

7.3.1 Mining

For mining operations, pollutants leakage and dewatering of the formation/deposit, as well as diversion of water flow are major issues related to water use. The removal of water from nearby aquifers can lower the overall water level in the area and may affect other aquifers and surface water bodies, including wetlands that are dependent on groundwater recharge.

The current method for the recovery of bitumen from the oil sands via surface mining results in the accumulation of large volumes of fluid wastes (*tailings*—a complex system of clays, minerals, and organic constituents), which are stored in large ponds until they can be used to begin filling in the mined out pits. The prevention of seepage of pollutants from tailings ponds, pits, and landfills into freshwater aquifers is an ongoing environmental concern (ERCB, 2008; NEB, 2006; Mikula et al., 2008).

Approved oil sands mining operations are currently licensed to divert water from the Athabasca River. A barrel of synthetic crude oil produced by mining operations, bitumen separation, and bitumen conversion needs 2.0–4.5 bbl of water. Approved oil sands mining operations will take more water from the Athabasca River—the current oil sands water license allocations are only for approximately 1% v/v of the flow of the river. The Alberta government sets strict limits on how much water oil sands companies can remove from the Athabasca River, and during low-flow conditions, orders them to reduce their withdrawals.

As noted above, tailings streams are by-products of the oil sands extraction process. Each ton of oil sand in place has a volume of about 16 ft^3, which will generate about 22 ft^3 of tailings giving a volume gain on the order of 40%. If an oil sand mine produces about 200,000 tons of oil sand per day, the volume expansion represents a considerable solids disposal problem. Tailings from the process consist of about 49% w/w sand, 1% w/w bitumen (organics), and approximately 50% w/w water.

After bitumen extraction, the tailings are pumped to a settling basin. Coarse tailings settle rapidly and can be restored to a dry surface for reclamation. Fine tailings, consisting of slow-settling clay particles and water, settle more slowly appearing to reach a point where settling time is difficult to measure.

7.3.2 *In Situ* Processes

For *in situ* processes, the water requirement to produce a barrel of recovered bitumen/oil with *in situ* (emphasis added) production may be less than 1 bbl (even as little as 0.2 bbl), depending on how much water is recycled. An *in situ* facility requires freshwater to generate steam, for various utility functions throughout the plant, separation of the bitumen from sand, hydrotransportation of bitumen slurry, and upgrading of the bitumen into lighter forms of oil for transport. However, as water is used to extract the bitumen in the ground, the *in situ* process has the negative effect of removing water permanently from the hydrologic cycle.

The demand for freshwater for *in situ* oil sands projects is projected to more than double to 82 million barrels by 2015 (NEB, 2006). In SAGD operations, 90–95% of the water used for steam to recover bitumen is reused, but for every 6.3 bbl of bitumen produced, approximately 1.3 bbl of additional groundwater must be used. Thus, even though the majority of the water is recycled in the SAGD process, the process still requires large volumes of water to be available.

There have been several initiatives to develop new technologies and integrated approaches to reduce water consumption. The development of a nonthermal *in situ* recovery method would reduce *in situ* water consumption significantly. Another challenge facing *in situ* operations is the potential for contamination of groundwater. Design improvements, monitoring, and surveillance systems may reduce the risk of damage to the aquifer and minimize the release of fluid to groundwater (NEB, 2006).

7.4 GREENHOUSE GASES

It has been estimated that for every barrel of synthetic oil produced at oil sand facilities in Alberta, more than 35 lb of greenhouse gases (GHGs) are released into the atmosphere and between 2 and 4 bbl of wastewater are sent to the tailings ponds, which are highly toxic (MacKinnon and Boerger, 1986).

Emissions of GHGs are one of the most complicated future environmental issues for the oil sands industry. Development of oil sand leases causes the emissions of carbon dioxide (CO_2), methane (CH_4), and nitrous oxide (N_2O). These are members of the group of *GHGs* that have the ability to affect the global climate. Oil sands operations include a wide spectrum of other air emissions such as (NEB, 2006):

- Sulfur dioxide (SO_2)
- Nitrogen oxides (NO_x)

- Hydrogen sulfide (H_2S)
- Carbon monoxide (CO)
- Volatile organic compounds (VOCs)
- Ozone (O_3)
- Polycyclic aromatic hydrocarbons (PAH)
- Particulate matter (PM)

A major Canadian initiative—the Integrated Carbon Dioxide Network (ICO_2N; http://www.ico2n.com/), whose members represent a group of industry participants providing a framework for carbon capture and storage development in Canada—has proposed a system for the large-scale capture, transport, and storage of carbon dioxide (CO_2).

7.5 THE FUTURE

There have been numerous forecasts of world production and demand for conventional crude, all covering varying periods of time, and even after considering the impact of the conservation ethic, the development of renewable resources, and the possibility of slower economic growth, nonconventional sources of liquid fuels could well be needed to make up for the future anticipated shortfalls in conventional supplies of crude oil.

At current rates of production, the Athabasca oil sands reserves as a whole could last over 400 years. New *in situ* methods have been developed to extract bitumen from deep deposits by injecting steam to heat the sands and reduce the bitumen viscosity so that it can be pumped out like conventional crude oil.

As the oil sands industry continues to expand, and other conventional sources of oil and gas are depleted, it is inevitable that oil sands will play a greater role in meeting energy needs. The oil sands companies are involved with ongoing research to make the oil sands plants run more efficiently. In addition new companies from Canada and other countries are showing a keen interest in the oil sands deposits of Northern Alberta.

Development of resources such as oil sand, oil shale, and coal is of particular interest to the United States, which has additional compelling reasons to develop viable alternate fossil fuel technologies. There has been the hope that the developing technology in the United States will eventually succeed in developing alternate energy sources. However, the optimism of the 1970s and 1980s has been succeeded by the reality of the twenty-first century and it is now obvious that these energy sources will not be the answer to energy shortfalls in the near term. Energy demands will most probably need to be met by the production of more liquid fuels from fossil fuel sources (Speight, 2011a,b, 2013).

There are those energy pundits who suggest that there is an inevitable decline of the *liquid fuel culture*. This is no secret nor is it rocket science—when a finite resource which took millions of years to form under geologic conditions is consumed there *has* to be a depletion of the resource. However, the potential for greater energy availability from alternative fossil fuel technologies is high. The United States is rich in coal, oil shale, and oil sand deposits, which (as a cumulative resource) are so extensive that with the development of appropriate technologies (and the support of various levels of government), the nation could be self-sufficient in energy by the turn of the twenty-first century.

Any degree of maximizing the production of liquid fuels will require the development of oil sand deposits. To develop the present concept of extracting oil from the oil sands, it is necessary to combine three operations, each of which contributes significantly to the cost of the venture: (1) a mining operation capable of handling two million tons, or more, of oil sand per day, (2) an extraction process to release bitumen from the sand, and (3) an upgrading plant to convert bitumen to synthetic crude oil.

However, in the United States, oil sand economics is still very much a matter for conjecture. The estimates published for current and proposed Canadian operations are, in a sense, not applicable to operations in the United States because of differences in the production techniques that may be required.

Finally, the fact that most of the oil sand resource in the United States is too deep for economic development is reflected in the ratio of the numbers of *in situ* projects to mining/extraction projects (almost 4:1).

Obviously, there are many features to consider when development of oil sand resources is planned. It is more important to recognize that what are important features for one resource might be less important in the development of a second resource. Recognition of this facet of oil sand development is a major benefit that will aid in the production of liquid fuels in an economic and effective manner.

REFERENCES

Charpentier, A.D., Bergerson, J.A., MacLean, H.L., 2009. Understanding the Canadian oil sands industry's greenhouse gas emissions. Environ. Res. Lett. 4 (1), 014005.

Dowdeswell, L., Dillon P., Ghoshal, S., Miall, A., Rasmussen, J., Smol, J.P., 2010. A Foundation for the Future: Building an Environmental Monitoring System for the Oil Sands. A Report Submitted to the Minister of Environment. Environment Canada, Ottawa, Ontario, Canada, December.

ERCB, 2008. ERCB Releases Draft Directive on Oil Sands Tailings Management and Enforcement Criteria. Alberta Energy Resources Conservation Board, Calgary, Alberta, Canada, June 26.

MacKinnon, M., Boerger, H., 1986. Description of two treatment methods for detoxifying oil sands tailings pond water. Water Pollut. Res. J. Can. 21, 496–512.

Mikula, R.J., Munoz, V.A., Omotoso, O., 2008. Water use in bitumen production: tailings management in surface mined oil sands. In: Proceedings of the World Heavy Oil Congress, Edmonton, Alberta, Canada, p. 1.

NEB, 2006. Canada's Oil Sands, Opportunities and Challenges to 2015: An Update. National Energy Board, Calgary, Alberta, Canada.

Speight, J.G., 2011a. The Refinery of the Future. Gulf Professional Publishing, Elsevier, Oxford.

Speight, J.G., 2011b. The Biofuels Handbook. Royal Society of Chemistry, London.

Speight, J.G., 2013. The Chemistry and Technology of Coal, third ed. CRC, Taylor & Francis Group, Boca Raton, FL.

Swart, N.C., Weaver, A.J., 2012. The Alberta oil sands and climate. Nat. Clim. Change 2 (3), 134–136.

GLOSSARY

Abandonment pressure: a direct function of the economic premises; it corresponds to the static bottom pressure at which the revenues obtained from the sales of the hydrocarbons produced are equal to the well's operation costs.

Absolute permeability: ability of a rock to conduct a fluid when only one fluid is present in the pores of the rock.

Acidizing: a technique for improving the permeability of a reservoir by injecting acid.

Air injection: an oil recovery technique using air to force oil from the reservoir into the wellbore.

Alberta: a province in western Canada—the capital is the city of Edmonton; the most populous city and metropolitan area, Calgary, is Alberta's economic hub and is located in the southern region of the province. To the west, Alberta's border with British Columbia follows the line of peaks of the Rocky Mountains range along the Continental Divide.

Alkaline flooding: *see* EOR process.

American Society for Testing and Materials (ASTM): the official organization in the United States for designing standard tests for petroleum and other industrial products.

Anticline: structural configuration of a package of folding rocks and in which the rocks are tilted in different directions from the crest.

API gravity: a measure of the *lightness* or *heaviness* of petroleum which is related to density and specific gravity. $°API = (141.5/\text{specific gravity at } 60°F) - 131.5$.

Apparent viscosity: the viscosity of a fluid, or several fluids flowing simultaneously, measured in a porous medium (rock), and subject to both viscosity and permeability effects; also called effective viscosity.

Asphaltene (asphaltene constituents): the brown to black powdery material produced by treatment of petroleum, petroleum residua, or bituminous materials with a low boiling liquid hydrocarbon, for example, pentane or heptane; soluble in benzene (and other aromatic solvents), carbon disulfide, and chloroform (or other chlorinated hydrocarbon solvents).

Barrel: the unit of measurement of liquids in the petroleum industry; equivalent to 42 US standard gallons or 33.6 imperial gallons.

Basin: receptacle in which a sedimentary column is deposited that shares a common tectonic history at various stratigraphic levels.

Billion: 1×10^9.

Bitumen: exists in oil sand (tar sand) deposits and is not recoverable by conventional secondary or tertiary methods; a naturally occurring viscous mixture of hydrocarbonaceous material which, in its naturally occurring state, is not recoverable by conventional recovery techniques.

Bituminous: containing bitumen or constituting the source of bitumen.

Bituminous rock: *see* Bituminous sand.

Bituminous sand: a formation in which the bituminous material (*see* Bitumen) is found as a filling in veins and fissures in fractured rock or impregnating relatively shallow sand, sandstone, and limestone strata; a sandstone reservoir that is impregnated with a heavy, viscous, black, petroleum-like material that cannot be retrieved through a well by conventional production techniques.

CFR: Code of Federal Regulations; Title 40 (40 CFR) contains the regulations for protection of the environment.

Chemical flooding: *see* Enhanced Oil Recovery (EOR) process.

Coal tar: the specific name for the tar (*q. v.*) produced from coal.

Coal tar pitch: the specific name for the pitch (*q. v.*) produced from coal.

COFCAW: an EOR process (*q. v.*) that combines forward combustion and waterflooding.

Cold production: the use of operating and specialized exploitation techniques in order to rapidly produce heavy oils without using thermal recovery methods.

Conventional crude oil (conventional petroleum): crude oil that is pumped from the ground and recovered using the energy inherent in the reservoir; also recoverable by application of secondary recovery techniques.

cP (centipoise): a unit of viscosity.

Cyclic steam injection: the alternating injection of steam and production of oil with condensed steam from the same well or wells.

Deasphaltened oil: the fraction of petroleum remaining after the asphaltene constituents have been removed.

Deasphaltening: removal of a solid powdery asphaltene fraction from petroleum by the addition of low boiling liquid hydrocarbons such as *n*-pentane or *n*-heptane under ambient conditions.

Deasphalting: the removal of the asphaltene fraction from petroleum by the addition of a low boiling hydrocarbon liquid such as *n*-pentane or *n*-heptane; more correctly the removal of asphalt (tacky, semisolid) from petroleum (as occurs in a refinery asphalt plant) by the addition of liquid propane or liquid butane under pressure.

Density: the mass (or weight) of a unit volume of any substance at a specified temperature; *see also* Specific gravity.

Deposit: a rock (such as sandstone)—technically, loose sand or partially consolidated sandstone containing naturally occurring mixtures of sand, clay, and water—which also contains bitumen.

Displacement efficiency: the ratio of the amount of oil moved from the zone swept by reprocessing to the amount of oil present in the zone prior to the start of the process.

Dome: geological structure with a semispherical shape or relief.

Downhole steam generator: a generator installed downhole in an oil well to which oxygen-rich air, fuel, and water are supplied for the purposes of generating steam for injection into the reservoir. Its major advantage over a surface steam generating facility is that the losses to the wellbore and surrounding formation are eliminated.

Dragline: a mining machine which drops a heavy toothed bucket on a cable from the end of a boom into the oil sand, then drags the bucket through the deposit, scooping up the sand. Once full, the bucket is raised and emptied into a windrow.

Dykstra–Parsons coefficient: an index of reservoir heterogeneity arising from permeability variation and stratification.

Effective permeability: a relative measure of the conductivity of a porous medium for a fluid when the medium is saturated with more than one fluid. This implies that the effective permeability is a property associated with each reservoir flow, for example, gas, oil, and water. A fundamental principle is that the total of the effective permeability is less than or equal to the absolute permeability.

Effective porosity: a fraction that is obtained by dividing the total volume of communicating pores and the total rock volume.

Effective viscosity: *see* Apparent viscosity.

Enhanced oil recovery (EOR) process: a method for recovering additional oil from a petroleum reservoir beyond that economically recoverable by conventional primary and secondary recovery methods. EOR methods are usually divided into three main categories: (1) *chemical flooding*: injection of water with added chemicals into a petroleum reservoir. The chemical processes include surfactant flooding, polymer flooding, and alkaline flooding; (2) *miscible flooding*: injection into a petroleum reservoir of a material that is miscible, or can become miscible, with the oil in the reservoir. Carbon dioxide, hydrocarbons, and nitrogen are used; and (3) *thermal recovery*: injection of steam into a petroleum reservoir, or propagation of a combustion zone through a reservoir by air or oxygen-enriched air injection. The thermal processes include steam drive, cyclic steam injection, and *in situ* combustion.

Expanding clays: clays that expand or swell on contact with water, for example, montmorillonite.

Extraction: the process by which the bitumen is separated from sand, water, and other impurities.

Extra heavy oil: a type of crude oil with properties similar to those of bitumen but having mobility in the reservoir.

FAST: fracture-assisted steam flood technology.

Fault: fractured surface of geological strata along which there has been differential movement. Fluid saturation: portion of the pore space occupied by a specific fluid; oil, gas, and water may exist.

FE 76-4: the definition of tar sand—the several rock types that contain an extremely viscous hydrocarbon which is not recoverable in its natural state by conventional oil well production methods including currently used enhanced recovery techniques.

Foamy oil: the oily product from a heavy oil reservoir when in solution—gas drive is used as the recovery method.

Flood, flooding: the process of displacing petroleum from a reservoir by the injection of fluids.

Formation: an interval of rock with distinguishable geologic characteristics.

Fractional composition: the composition of petroleum as determined by fractionation (separation) methods.

Froth: a mixture of air, water, and bitumen which rises to the surface of the primary separation vessel.

Gravity: *see* API gravity.

Gravity drainage: the movement of oil in a reservoir that results from the force of gravity.

Gravity segregation: partial separation of fluids in a reservoir caused by gravitational force acting on differences in density.

Gravity-stable displacement: the displacement of oil from a reservoir by a fluid of a different density, where the density difference is utilized to prevent gravity segregation of the injected fluid.

Heavy oil: petroleum typically having an API gravity of less than 20° which is recoverable from the reservoir using conventional recovery methods.

Heavy petroleum: *see* Heavy oil.

Heterogeneity: lack of uniformity in reservoir properties such as permeability.

Horizontal sweep efficiency: the fraction of the layers or horizontal zones of a reservoir that are effectively contacted by displacing fluids (see Vertical; sweep efficiency).

Hot production: the optimum production of heavy oil through use of enhanced thermal recovery methods.

Huff and puff: a cyclic EOR method in which steam or gas is injected into a production well; after a short shut-in period, oil and the injected fluid are produced through the same well.

Hybrid process: a combination of any existing or new technologies used for *in situ* production of bitumen. The adoption of hybrid processes by the oil sands industry may allow for more efficient use of energy and help to reduce environmental impacts.

Hydraulic fracturing: the opening of fractures in a reservoir by high-pressure, high-volume injection of liquids through an injection well.

Hydrocarbon-producing resource: a resource such as coal and oil shale (kerogen) which produces derived hydrocarbons by the application of conversion processes; the hydrocarbons so-produced are not naturally occurring materials.

Hydrocarbon resource: a resource such as petroleum or natural gas which can produce naturally occurring hydrocarbons without the application of conversion processes.

Immiscible: two or more fluids not having complete mutual solubility and coexisting as separate phases.

Immiscible carbon dioxide displacement: injection of carbon dioxide into an oil reservoir to effect oil displacement under conditions in which miscibility with reservoir oil is not obtained.

Immiscible displacement: a displacement of oil by a fluid (gas or water) that is conducted under conditions so that interfaces exist between the driving fluid and the oil.

Incremental ultimate recovery: the difference between the quantity of oil that can be recovered by EOR methods and the quantity of oil that can be recovered by conventional recovery methods.

Injection well: a well in an oil field used for injecting fluids into a reservoir.

Injectivity: the relative ease with which a fluid is injected into a porous rock.

In situ: in its original place; in the reservoir.

In situ mining: a nonmining process for recovering bitumen from deep oil sand deposits.

In situ **combustion:** an EOR process consisting of injecting air or oxygen-enriched air into a reservoir under conditions that favor burning part of the *in situ* petroleum, advancing this burning zone, and recovering oil heated from a nearby producing well.

Kinematic viscosity: the ratio of viscosity (*q.v.*) to density, both measured at the same temperature.

Light crude oil (light petroleum, conventional crude oil): oil having an API gravity of at least 20° and a viscosity less than 100 centipoise (cP).

Lithology: the geological characteristics of reservoir rock.

Maltenes: that fraction of petroleum that is soluble in, for example, pentane or heptane; deasphaltened oil (*q.v.*); also the term arbitrarily assigned to the pentane-soluble portion of petroleum that is relatively high boiling (>300°C, 760 mm).

MEOR: microbial EOR.

Micellar fluid (surfactant slug): an aqueous mixture of surfactants, cosurfactants, salts, and hydrocarbons. The term micellar is derived from the word micelle, which is a submicroscopic aggregate of surfactant molecules and associated fluid.

Microemulsion: a stable, finely dispersed mixture of oil, water, and chemicals (surfactants and alcohols).

Microemulsion or micellar/emulsion flooding: an augmented waterflooding technique in which a surfactant system is injected in order to enhance oil displacement toward producing wells.

Microorganisms: animals or plants of microscopic size, such as bacteria.

Microscopic displacement efficiency: the efficiency with which an oil displacement process removes the oil from individual pores in the rock.

Middle-phase microemulsion: a microemulsion phase containing a high concentration of both oil and water that, when viewed in a test tube, resides in the middle with the oil phase above it and the water phase below it.

Mineable oil sand: oil sand which can be recovered by surface mining.

Minerals: naturally occurring inorganic solids with well-defined crystalline structures.

Miscibility: an equilibrium condition, achieved after mixing two or more fluids, which is characterized by the absence of interfaces between the fluids: (1) *first-contact miscibility*: miscibility in the usual sense, whereby two fluids can be mixed in all proportions without any interfaces forming. Example: At room temperature and pressure, ethyl alcohol and water are first-contact miscible. (2) *multiple-contact miscibility (dynamic miscibility)*: miscibility that is developed by repeated enrichment of one fluid phase with components from a second fluid phase with which it comes into contact. (3) *minimum miscibility* pressure: the minimum pressure above which two fluids become miscible at a given temperature, or can become miscible, by dynamic processes.

Miscible flooding: *see* EOR process.

Miscible fluid displacement (miscible displacement): an oil displacement process in which an alcohol, a refined hydrocarbon, a condensed petroleum gas, carbon dioxide, liquefied natural gas, or even exhaust gas is injected into an oil reservoir, at pressure levels such that the injected gas or fluid and reservoir oil are miscible; the process may include the concurrent, alternating, or subsequent injection of water.

Mobility: a measure of the ease with which a fluid moves through reservoir rock; the ratio of rock permeability to apparent fluid viscosity.

Mobility ratio: the mobility of the displacing phase divided by the mobility of the displaced phase.

Muskeg: a water soaked form of peat, sphagnum moss, 1–2 m thick, found on top of an overburden.

Native asphalt: *see* Bitumen.

Nonionic surfactant: a surfactant molecule containing no ionic charge.

Oil originally in place (OOIP): the quantity of petroleum existing in a reservoir before oil recovery operations begin.

Oil sand: a formation in which the bituminous material (bitumen) is found as a filling in veins and fissures in fractured rocks or impregnating relatively shallow sand, sandstone, and limestone strata; a sandstone reservoir that is impregnated with a heavy, extremely viscous, black hydrocarbonaceous, petroleum-like material that cannot be retrieved through a well by conventional or EOR techniques.

OOIP: *see* Oil originally in place.

Original oil volume in place: the amount of petroleum that is estimated to exist initially in the reservoir and that is confined by geologic and fluid boundaries, which may be expressed at reservoir or atmospheric conditions.

Original pressure: pressure prevailing in a reservoir that has never been produced. It is the pressure measured by a discovery well in a producing structure.

Overburden: the layer of sand, gravel, and shale which overlies the oil sands.

Override: the gravity-induced flow of a lighter fluid in a reservoir above another heavier fluid.

Pay zone thickness: the depth of an oil sand deposit from which bitumen (or a product) can be recovered.

Permeability: rock property for permitting a fluid pass. It is a factor that indicates whether a reservoir has producing characteristics or not.

Petroleum: mixture of hydrocarbons composed of combinations of carbon and hydrogen atoms found in the porous spaces of rocks. Crude oil may contain other elements of a nonmetal origin, such as sulfur, oxygen, and nitrogen, in addition to trace metals as minor constituents. The compounds that form petroleum may be in a gaseous, liquid, or solid state, depending on their nature and the existing pressure and temperature conditions.

Phase: a separate fluid that coexists with other fluids; gas, oil, water, and other stable fluids such as microemulsions are all called phases in EOR research.

Pitch: the nonvolatile, brown to black, semisolid to solid viscous product from the destructive distillation of many bituminous or other organic materials, especially coal.

Play: a group of fields that share geological similarities in which the geological trap controls the distribution of oil and gas.

Polymer: in EOR, any very high-molecular-weight material that is added to water to increase viscosity for polymer flooding.

Polymer-augmented waterflooding: waterflooding in which organic polymers are injected with the water to improve horizontal and vertical sweep efficiencies.

Pore volume: the total volume of all pores and fractures in a reservoir or part of a reservoir; also applied to catalyst samples.

Porosity: the ratio between the pore volume existing in a rock and the total rock volume; a measure of rock's storage capacity; the percentage of rock volume available to contain water or other fluid.

Pour point: the lowest temperature at which oil will pour or flow when it is chilled without disturbance under definite conditions.

Pressure pulse test: a technique for determining reservoir characteristics by injecting a sharp pulse of pressure in one well and detecting it in surrounding wells.

Pressure transient testing: measuring the effect of changes in pressure at one well on other wells in a field.

Primary recovery: the extraction of petroleum by only using the natural energy available in the reservoirs to displace fluids through the reservoir rock to the wells.

Producing well: a well in an oil field used for removing fluids from a reservoir.

Quadrillion: 1×10^{15}.

Recovery factor (rf): the ratio between the original volume of oil or gas, at atmospheric conditions, and the original reserves of the reservoir.

Relative permeability: the capacity of a fluid, such as water, gas, or oil, to flow through a rock when it is saturated with two or more fluids. The value of the permeability of a saturated rock with two or more fluids is different to the permeability value of the same rock saturated with just one fluid.

Reservoir: the portion of the geological trap containing petroleum or heavy oil that acts as a hydraulically interconnected system, and where the hydrocarbons are found at an elevated temperature and pressure occupying the porous spaces.

Retention: the loss of chemical components due to adsorption onto the rock's surface, to precipitation, or to trapping within the reservoir.

Rock matrix: the granular structure of a rock or porous medium.

Sand: a course granular mineral mainly comprising quartz grains that is derived from the chemical and physical weathering of rocks rich in quartz, notably sandstone and granite.

Sandstone: a sedimentary rock formed by compaction and cementation of sand grains; can be classified according to the mineral composition of the sand and cement.

Screening guide: a list of reservoir rock and fluid properties critical to an EOR process.

Secondary recovery: techniques used for the additional extraction of petroleum after primary recovery. This includes gas or water injection, partly to maintain reservoir pressure.

Sediment: an insoluble solid formed as a result of the storage instability and/or the thermal instability of petroleum and petroleum products.

Sedimentary strata: typically consist of mixtures of clay, silt, sand, organic matter, and various minerals; formed by or from deposits of sediments, especially from sand grains or silts transported from their source and deposited in water, such as sandstone and shale; or from calcareous remains of organisms, such as limestone.

Solvent-assisted process: any process that uses a low boiling hydrocarbon, such as propane, butane, or condensates, in the recovery of bitumen.

Sour crude oil: crude oil containing an abnormally large amount of sulfur compounds; *see also* Sweet crude oil.

Specific gravity: an intensive property of matter that is related to the mass of a substance and its volume through the coefficient between these two quantities. It is expressed in grams per cubic centimeter or in pounds per gallon.

Steam distillation: distillation in which vaporization of the volatile constituents is effected at a lower temperature by introduction of steam (open steam) directly into the charge.

Steam drive injection (steam injection): EOR process in which steam is continuously injected into one set of wells (injection wells) or other injection source to effect oil displacement toward and production from a second set of wells (production wells); steam stimulation of production wells is *direct steam stimulation*, whereas steam drive by steam injection to increase production from other wells is *indirect steam stimulation*.

Steam stimulation: injection of steam into a well and the subsequent production of oil from the same well.

Strata: layers including the solid iron-rich inner core, molten outer core, mantle, and crust of the earth.

Stratigraphy: part of geology that studies the origin, composition, distribution, and succession of rock strata.

Sucker rod pumping system: a method of artificial lift in which a subsurface pump located at or near the bottom of the well and connected to a string of sucker rods is used to lift the well fluid to the surface.

Surface-active material: a chemical compound, molecule, or aggregate of molecules with physical properties that cause it to adsorb at the interface between *two* immiscible liquids, resulting in a reduction of interfacial tension or the formation of a microemulsion.

Surface footprint: the amount of land disturbed by a development. Horizontal wells, multiwell pads, and progressive reclamation all help to minimize the surface footprint of *in situ* developments.

Surfactant: a type of chemical, characterized as one that reduces interfacial resistance to mixing between oil and water or changes the degree to which water wets reservoir rock.

Sweep efficiency: the ratio of the pore volume of reservoir rock contacted by injected fluids to the total pore volume of reservoir rock in the project area; (*see also* Horizontal sweep efficiency and Vertical sweep efficiency.)

Sweet crude oil: crude oil containing little sulfur; *see also* Sour crude oil.

Swelling: the increase in the volume of crude oil caused by absorption of EOR fluids, especially carbon dioxide. Also the increase in volume of clays when exposed to brine.

Swept zone: the volume of rock that is effectively swept by injected fluids.

Synthetic crude oil (syncrude): a hydrocarbon product produced by the conversion of oil sand bitumen and which resembles conventional crude oil.

Tailings: waste products from the mining, extraction, and upgrading process.

Tailings pond: an enclosure to contain tailings.

Tar: the volatile, brown to black, oily, viscous product from the destructive distillation of many bituminous or other organic materials, especially coal; a name used for petroleum in ancient texts.

Tar sand: *see* Oil sand; *see* Bituminous sand.

Thermal recovery: *see* EOR process.

Total thickness (*h*): thickness from the top of the formation of interest down to a vertical boundary determined by a water level or by a change of formation.

Trap: geometry that permits the concentration of hydrocarbons; a sediment in which oil and gas accumulate from which further migration is prevented.

Trillion: 1×10^{12}.

Ultimate analysis: elemental composition.

Ultimate recovery: the cumulative quantity of oil that will be recovered when revenues from further production no longer justify the costs of the additional production.

Upper-phase microemulsion: a microemulsion phase containing a high concentration of oil that, when viewed in a test tube, resides on top of a water phase.

Vertical sweep efficiency: the fraction of the layers or vertical zones of a reservoir that are effectively contacted by displacing fluids (see Horizontal sweep efficiency).

Visbreaking: a process for reducing the viscosity of heavy feedstocks by controlled thermal decomposition.

Viscosity: a measure of the ability of a liquid to flow or a measure of its resistance to flow; the force required to move a plane surface of area $1 \, m^2$ over another parallel plane surface $1 \, m$ away at a rate of $1 \, m/s$ when both surfaces are immersed in the fluid.

Waterflood: injection of water to displace oil from a reservoir (usually a secondary recovery process).

Waterflood mobility ratio: mobility ratio of water displacing oil during waterflooding; *see also* Mobility ratio.

Waterflood residual: the waterflood residual oil saturation; the saturation of oil remaining after waterflooding in those regions of the reservoir that have been thoroughly contacted by water.

Well abandonment: the final activity in the operation of a well when it is permanently closed under safety and environment preservation conditions.

Wellbore: the hole in the earth comprising a well.

Well completion: the complete outfitting of an oil well for either oil production or fluid injection; also the technique used to control fluid communication with the reservoir.

Wellhead: that portion of an oil well above the surface of the ground.

Wettability: the relative degree to which a fluid will spread on (or coat) a solid surface in the presence of other immiscible fluids.

Wettability number: a measure of the degree to which a reservoir rock is water wet or oil wet, based on capillary pressure curves.

Wettability reversal: the reversal of the preferred fluid wettability of a rock, for example, from water wet to oil wet, or vice versa.

CPSIA information can be obtained at www.ICGtesting.com
Printed in the USA
BVOW042334241012

303836BV00006B/57/P